Springer Theses

Recognizing Outstanding Ph.D. Research

Aims and Scope

The series "Springer Theses" brings together a selection of the very best Ph.D. theses from around the world and across the physical sciences. Nominated and endorsed by two recognized specialists, each published volume has been selected for its scientific excellence and the high impact of its contents for the pertinent field of research. For greater accessibility to non-specialists, the published versions include an extended introduction, as well as a foreword by the student's supervisor explaining the special relevance of the work for the field. As a whole, the series will provide a valuable resource both for newcomers to the research fields described, and for other scientists seeking detailed background information on special questions. Finally, it provides an accredited documentation of the valuable contributions made by today's younger generation of scientists.

Theses are accepted into the series by invited nomination only and must fulfill all of the following criteria

- They must be written in good English.
- The topic should fall within the confines of Chemistry, Physics, Earth Sciences, Engineering and related interdisciplinary fields such as Materials, Nanoscience, Chemical Engineering, Complex Systems and Biophysics.
- The work reported in the thesis must represent a significant scientific advance.
- If the thesis includes previously published material, permission to reproduce this must be gained from the respective copyright holder.
- They must have been examined and passed during the 12 months prior to nomination.
- Each thesis should include a foreword by the supervisor outlining the significance of its content.
- The theses should have a clearly defined structure including an introduction accessible to scientists not expert in that particular field.

More information about this series at http://www.springer.com/series/8790

Haoran Wen

Toward Inertial-Navigation-on-Chip

The Physics and Performance Scaling of Multi-Degree-of-Freedom Resonant MEMS Gyroscopes

Doctoral Thesis accepted by Georgia Institute of Technology, Atlanta, GA, USA

Springer

Haoran Wen
School of Electrical and Computer Engineering
Georgia Institute of Technology
Atlanta, GA, USA

ISSN 2190-5053 ISSN 2190-5061 (electronic)
Springer Theses
ISBN 978-3-030-25469-8 ISBN 978-3-030-25470-4 (eBook)
https://doi.org/10.1007/978-3-030-25470-4

This Springer imprint is published by the registered company Springer Nature Switzerland AG
The registered company address is: Gewerbestrasse 11, 6330 Cham, Switzerland

To my loving family

Supervisor's Foreword

Haoran Wen's thesis focuses on performance scaling of Coriolis vibratory gyroscopes and development of next-generation multi-axis gyroscopes using microelectromechanical systems (MEMS) technology for high-performance single-chip inertial measurement units (IMU). It covers a comprehensive study of the physics of resonator gyroscopes as well as novel engineering and microfabrication solutions to key performance limits in MEMS resonator gyroscopes, setting it apart from other theses in this area.

MEMS IMUs have been widely used in consumer electronics and automotive applications for motion detection and safety control. Improving their performance, robustness, and power consumption can broaden their uses to more advanced applications such as health informatics and indoor navigation. Conventional MEMS gyroscopes are limited by trade-offs between their performance and their power consumption and robustness. This thesis covers a thorough study of the physics of vibratory gyroscopes and provides new insight into the error sources and performance scaling of MEMS vibratory gyroscopes to break through the trade-offs for next-generation MEMS gyroscopes and IMUs. Theoretical, numerical, and experimental studies are well documented in this thesis, making it a valuable reference for future researchers in this field. Based on the theoretical studies, novel designs and fabrication platform are developed, featuring highly efficient nano-gap transduction and quadrature-error compensation that enable high-frequency resonant-type triaxial gyroscopes on a chip with high sensitivities and nanometer-scale actuations. In addition, this thesis also addressed other critical performance limits in MEMS gyroscopes such as nonlinearity and energy dissipation. Altogether, it created wafer-level-packaged single-chip IMU prototypes with co-fabricated high-frequency triaxial gyroscopes and high-bandwidth triaxial accelerometers. The prototypes demonstrated performance among the best for the IMUs to date while possessing unmatched robustness. The small size and low actuation of the gyroscopes also offer the potentials for robust flexible substrate integration and low-power operation, showing a path toward the next-generation IMUs for future low-power applications including wearable health informatics, personal navigation, and other advanced harsh-environment industrial applications.

In summary, Haoran's dissertation work provided deeper understandings of the physics of MEMS gyroscopes, offered innovative solutions to some of the most intriguing problems, and led to a unique enabling MEMS fabrication platform. I believe all of the above have significant impacts on developing the next-generation MEMS IMUs and can be of great value to researchers in the field.

Georgia Institute of Technology
Atlanta, GA, USA
June 2019

Farrokh Ayazi

Acknowledgments

First, I would like to express my sincere appreciation and gratitude to my advisor and mentor, Professor Farrokh Ayazi, not only for his support and guidance throughout my Ph.D. study but also for giving me the chance to develop my research in a field I like. I also want to thank him for placing his trust and confidence in me with various opportunities, which allowed me to gain experience and skills that are invaluable for my future career. I would like to thank my dissertation committee members, Professor Chandra Raman, Professor Peter J. Hesketh, Professor Phillip N. First, and Professor Simon Sponberg for their interest in my research and for spending their precious time on reviewing this dissertation and providing valuable feedback.

It has been a great pleasure and honor to work with and become friends with many former and current members of the IMEMS group. I would like to thank Dr. Diego Serrano, Dr. Roozbeh Tabrizian, Dr. Arashk Shirazi, and Dr. Yaesuk Jeong for all their invaluable discussions and suggestions and Pantani Liu, Sam Wisher, and Ryan Lei for their collaborative help. I give my special thanks to Anosh Daruwalla, Benoit Hamelin, and Mojtaba Hodjat-Shamami. Together, we survived cleanroom. In addition, I appreciate all the helps provided by the cleanroom staffs at Georgia Institute of Technology, which made this work possible.

I want to express my deepest gratitude to my parents, for their unconditional love and support have given me the courage and confidence to chase my dreams. Last but not least, I wish to thank Dongxiunan Mao for her support and patience during my graduate studies and for always being there for me to keep me positive and motivated.

Contents

Parts of this thesis have been published in the following articles:

Journal articles:

- H. Wen, A. Daruwalla, C.S. Liu, F. Ayazi, A high-frequency resonant framed-annulus pitch or roll gyroscope for robust high-performance single-chip inertial measurement units. J. Microelectromech. Syst. **27**(6), 995–1008 (2018)
- H. Wen, A. Daruwalla, F. Ayazi, Resonant pitch and roll silicon gyroscopes with sub-micron-gap slanted electrodes: Breaking the barrier towards high-performance monolithic inertial measurement units. Microsyst. Nanoeng. **3**, 16092 (2017).

Conference proceedings:

- H. Wen, A. Daruwalla, Y. Jeong, P. Gupta, J. Choi, C.S. Liu, F. Ayazi, Wafer-level-packaged HARPSS+ MEMS platform: integration of robust timing and inertial measurement units (TIMU) on a single chip, in *Proceedings of 2018 IEEE/ION Position, Location and Navigation Symposium (PLANS 2018)*, Monterey, Apr 2018, pp. 261–266
- H. Wen, A. Daruwalla, Y. Jeong, P. Gupta, J. Choi, C.S. Liu, F. Ayazi, A high-performance single-chip timing and inertial measurement unit with robust mode-matched gyroscopes, in *Proceedings of the 31st IEEE International Micro Electro Mechanical Systems Conference (MEMS 2018)*, Belfast, Jan 2018, pp. 105–108
- H. Wen, A. Daruwalla, F. Ayazi, A 0.5MHz mode-matched pitch or roll annulus gyroscope with nano-gap slanted electrodes for quadrature cancellation, in *Tech Digest of the 19th International Conference on Solid- State Sensors, Actuators and Microsystems (TRANSDUCERS'17)*, Kaohsiung, June 2017, pp. 536–539
- H. Wen, S. Wisher, F. Ayazi, An anisotropic-wet-etched pitch or roll mode-matched gyroscope with slanted quadrature-cancellation electrodes, in *Tech Digest of 2016 Solid-State Sensors, Actuators, and Microsystems Workshop (Hilton Head 2016)*, Hilton Head Island, June 2016, pp. 226–229
- H. Wen, A. Daruwalla, R. Mirjalili, F. Ayazi, HARPSS-fabricated nano-gap comb-drive for efficient linear actuation of high frequency BAW resonators, in *Proceedings of the 29th IEEE International Conference on Micro Electro Mechanical Systems (MEMS 2016)*, Shanghai, Jan 2016, pp. 1014–1017
- R. Mirjalili, H. Wen, D.E. Serrano, F. Ayazi, Substrate-decoupled silicon disk resonators having degenerate gyroscopic modes with Q in excess of 1-million, in *Tech Digest of the 18th International Conference on Solid-State Sensors, Actuators and Microsystems (TRANSDUCERS'15)*, Anchorage, June 2015, pp. 15–18

Chapter 1
Introduction

From the ancient celestial method to the modern Global Positioning System (GPS), navigation technologies have been playing an important role in shaping the way people understand and experience the world. And efforts have never stopped in advancing the navigation technologies to bring more enabling direction guidance and to allow exploration of more unknown territories [1–3].

The GPS first became operational in the 1990s; since then, GPS devices have made traveling much more convenient and efficient. However, relying on external satellite reference signals puts a fundamental limit on their applications, and people have been seeking portable solutions for navigational applications in environments with interrupted or no satellite signals. For example, indoor navigation equipment with small size, weight, power, and cost (SWAP-C) has great values in realizing applications like in-residence guidance for visually impaired person [4] and automatic search-and-rescue robots [5]. In addition, miniaturized satellite-free navigation devices have many scientific applications such as tracking wild animals in dense forests [6] and providing guidance for individuals during underwater or space exploration missions [7]. Besides, portable navigational grade sensors may enable experiments for fundamental physics studies like earth rotation measurement [8] as well as other geophysical and relativity tests [9, 10]. Therefore, developing reliable stand-alone navigation technology with small size and low cost is highly desirable for both improving daily life experience and advancing scientific studies.

1.1 Inertial Navigation System

Unlike GPS, INSs are self-contained systems providing independent position tracking [11, 12]. The elimination of reference signals makes INS more robust and flexible compared to GPS and other technologies relying on reference objects. An INS is enabled by a high-precision IMU including accelerometers for 3-axis (x, y, z) linear motion detection and gyroscopes for 3-axis (pitch, roll, yaw) rotation

H. Wen, *Toward Inertial-Navigation-on-Chip*, Springer Theses,
https://doi.org/10.1007/978-3-030-25470-4_1

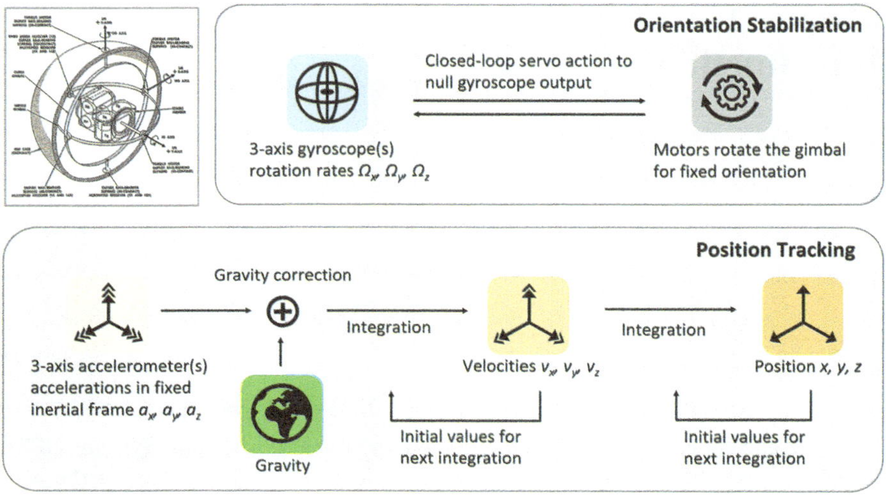

Fig. 1.1 Schematic of the operation of a stable-platform INS. Top-left inset shows IMU of Lunar Modules used in the Apollo Missions [19]

detection. INSs can be implemented with IMUs in either a stable-platform or a strapdown configuration. In a stable-platform INS, 3-axis accelerometers and 3-axis gyroscopes are mounted on a platform levitated with motor controlled gimbal structures [13]. The gyroscopes and the servo motors form a closed loop where the motors continuously react to and null changes in gyroscope output signal to maintain the platform at a constant orientation with respect to a fixed inertial frame. The accelerometers measure the linear acceleration along the three axes of the fixed coordinate system, and the position is calculated through a double integration of the accelerations (Fig. 1.1). Precisely machined stable-platform IMUs can provide supreme accuracy and robustness, and they have been used since the 1950s and are still in use nowadays for aerospace navigation applications. However, the high performance of stable-platform IMUs comes at the price of highly complicated gimbal structures and motors with large weight and volume and requiring expensive high-precision machining techniques. Those limits make stable-platform IMUs not feasible for individual or personal applications requiring low cost and small form factor [14, 15].

On the contrary, a strapdown system has sensors integrated on a platform that is fixed to the object under measurement [16]. In a strapdown system, rotation rates measured by the gyroscopes are integrated to determine the instant angles of the platform with respect to a fixed inertial frame. The orientation of the platform defines a rotated coordinate system, and linear accelerations along the axes of the rotated coordinate system are captured by the accelerometers. The acceleration vector in the rotated frame is projected to the fixed frame using the rotation angles obtained from gyroscope measurements. And the position of the measured object is determined by double integration of the projected accelerations (Fig. 1.2). The fixed configuration

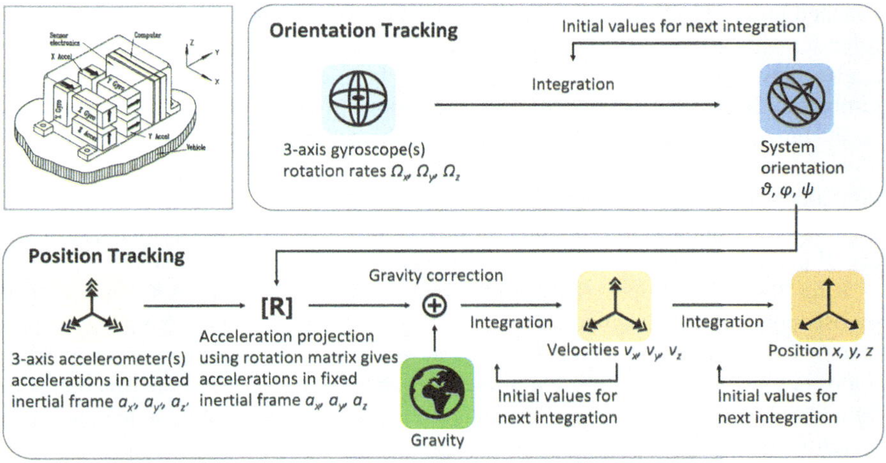

Fig. 1.2 Schematic of a strapdown INS [20] and the operation principle

of a strapdown system eliminates the need for motored gimbal structures, allowing the IMUs to have more compact form, lower fabrication cost, and lower power consumption. The best examples to show such advantages are MEMS IMUs [17, 18]. Benefiting from the fast development of micromachining technologies, low-cost batch fabrication of silicon MEMS IMUs is possible with multi-DOF sensors integrated on a single chip. The small size and light weight of MEMS IMUs make them ideal for hand-held devices and wearable electronics. However, a drawback of strapdown system is that the position tracking is tightly related to the orientation tracking. Unlike in stable-platform system where rotation angles are never measured, the rotation angles of a strapdown system need to be evaluated constantly by integrating the gyroscope output. As a result, strapdown systems are vulnerable to gyroscope error accumulation and propagation. In order to build strapdown INSs for personal applications, developing miniaturized high-performance gyroscopes is the most crucial task.

1.2 Evolution of Gyroscopes

Gyroscopes are physical sensors measuring the rotation of an object relative to an inertial reference frame. Based on the operation principle, gyroscopes can be divided into three categories: mechanical gyroscopes, optical gyroscopes, and atomic gyroscopes.

Traditional mechanical gyroscopes are gimballed spinning wheel gyroscopes [21]. The spinning wheel is called rotor, which is mounted on a frame through a set of gimbals that ideally allows the rotor to rotate freely around three axes. When attached to an object, the frame rotates with the object, while the orientation of the rotor remains fixed due to the conservation of angular momentum. The relative angles between the rotor and the frame can be measured at the bearings of the gimbals to directly determine the angle of rotations of the object under measurement. As it directly measures the rotation angles, there are no accumulative errors and high-precession measurements can be obtained. However, the disadvantage of gimballed spinning wheel gyroscopes is that friction of the gimbals will cause precession that appears as orientation drift, and minimizing the drift requires the use of precision mechanical moving parts that are bulky and expensive.

Ring laser gyroscopes (RLG) and fiber optic gyroscopes (FOG) are optical gyroscopes based on Sagnac effect [22]. Sagnac effect describes the phenomenon that two light beams with the same frequency traveling around a closed path in opposite directions have different phase shift under rotation due to the invariance of light speed in all inertial frames. The interference pattern of such two light beams can therefore be used as a measurement of the rotation. Optical gyroscopes have no moving parts and can provide high-accuracy rotation measurement. They are widely used in aerospace and marine applications for navigational guidance. Large RLGs have also been found useful for earth rotation and gravitational wave detection applications. Despite their extremely high accuracy and reliability, both RLG and FOG rely on large effective area of the closed path and high-precision optical components to achieve high performance, which makes their miniaturization extremely challenging and prohibits them from applications using portable or hand-held devices (Fig. 1.3).

Recent developments in modern physics have enabled a new group of gyroscopes based on atomics phenomenon, including the atomic interferometer gyroscope (AIG) and atomic spin gyroscope (ASG) [25]. The AIG is based on a principle like the Sagnac effect using atoms instead of photons, whereas ASG is based on the nuclear magnetic resonance (NMR) phenomena of atomic spin. Both technologies are aiming for ultra-high-precision rotation sensing. However, the manipulation of atoms and control of temperature and magnetic field involve expensive and bulky equipment and long start-up time. Although efforts are being made in miniaturizing

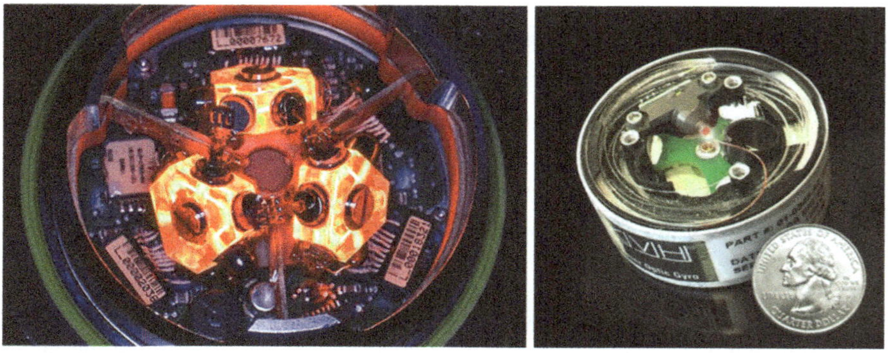

Fig. 1.3 Honeywell RLG [23] and KVH DSP-1750 FOG [24]

ASG, performance achievable with miniature ASG is yet to match the values promised by the technology, and they are also still a long way from meeting the low SWAP-C requirements for personal applications (Fig. 1.4).

Another class of mechanical gyroscopes is Coriolis vibratory gyroscopes (CVG) [27]. Instead of using gimballed moving parts, vibrating structures are used as sensing elements. And instead of using the conservation of angular momentum, they rely on the vibrating element's response to a rotation-induced inertial force called Coriolis force. Details of the operation principle will be discussed in the next chapter. CVG not only allows a smaller form factor but also eliminates problems of friction and wear in traditional mechanical gyroscopes. A successful example of CVG is quartz hemispherical resonator gyroscope (HRG) [28]. HRGs can deliver extremely high reliability with very low power dissipation, but they require expensive high-precision manufacture which limits them to mostly aerospace applications (Fig. 1.5).

Fig. 1.4 Northrop Grumman micronuclear magnetic resonance gyro [26]

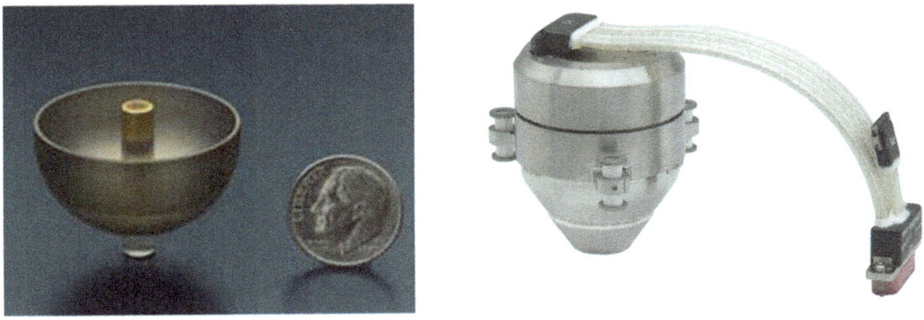

Fig. 1.5 Northrop Grumman HRG resonator and sensor assembly [28]

Over the past two decades, significant progresses have been made in developing MEMS CVG owing to advancements in low-cost batch fabrication silicon micromachining technologies [29]. MEMS CVG has demonstrated incomparable capability in reaching low SWAP-C and allowing multi-DOF sensor fusion, which has brought it great success in consumer, automotive, and industrial markets. Although MEMS gyroscopes currently are not delivering performance as good as conventional mechanical and optical gyroscopes, they are far from reaching their full potentials. Considering the unbeatable advantages of low cost, small size, and high integratability of MEMS gyroscopes, the most promising way to enable emerging portable INS applications is by understanding the physics behind MEMS gyroscopes and improving the performance of multi-DOF MEMS gyroscopes to enable IMUs for short-term and long-term navigational tasks, which is the aim of this work.

1.3 Evolution of MEMS Gyroscopes

The first silicon MEMS gyroscopes were introduced in the 1990s [30, 31]. And driven by a large market demand from automotive safety control applications, great amount efforts were put into improvement and commercialization of MEMS gyroscopes in the early 2000s. The first commercial MEMS yaw gyroscope came from ADI in 2002 [32], and pitch and roll gyroscopes were first brought into market by InvenSense in 2006 [33] (Fig. 1.6).

Later on, the blossom of personal electronics again brought a huge costumer market for MEMS gyroscopes and boosted their development in the direction of low-cost compact integrated multi-axis gyroscope units (Fig. 1.7).

To date, most commercial MEMS gyroscopes are tuning fork gyroscopes (TFG) or variations utilizing similar operational motions. For yaw TFG, a pair of proofmass is driven to vibrate in x (or y) direction, when being rotated around z-axis, the proofmass experiences Coriolis force in y (or x) direction and undergoes movement

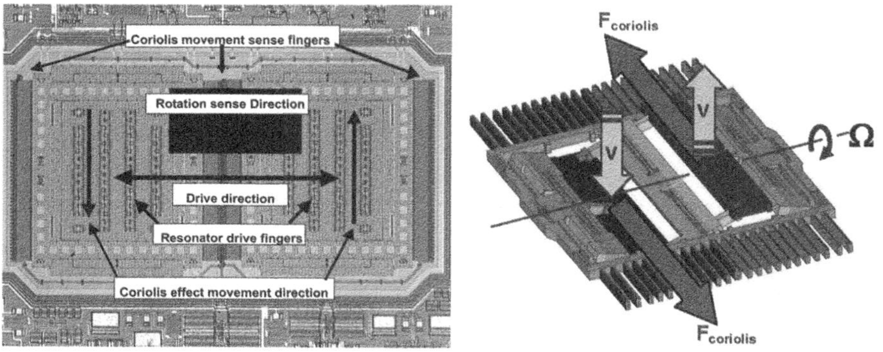

Fig. 1.6 (Left) Mechanical structure and operational movements of the first commercial ADI yaw gyroscope. (Right) operational movement of InvenSense roll gyroscope [33]

Fig. 1.7 Infrared image of InvenSense MPU-6500 6-DOF IMU with package size of 3 × 3 × 0.9 mm^3 [34]

Table 1.1 Gyroscope performance classification [18]

Parameter	Rate grade	Tactical grade	Inertial grade
Full-scale range (°/s)	50–1000	>500	>400
Bandwidth (Hz)	>70	~100	~100
Angle random walk (°/√h)	> 0.5	0.5–0.05	<0.001
Bias instability (°/h)	10–1000	0.1–10	<0.01

in *y* (or *x*) direction accordingly which is detected to determine the rotation rate. Pitch or roll gyroscopes are implemented in a similarly fashion, with proper exchanges of driving, detecting, and rotation axis. Over the years, progresses are being made to reduce the size of commercial multi-axis gyroscope units while gradually improving their resolutions from rate grade toward tactical grade (Table 1.1).

However, TFGs use large proofmass and drive amplitudes to achieve large sensitivity-to-noise ratio and reduce ARW for high resolutions. Therefore, there is a trade-off between the performance and form factor of the TFGs, which puts a fundamental limit on the achievable performance with a certain device size. In addition, using large drive amplitude with low power consumption requires the vibrating structure to be flexible with low stiffness. The low stiffness combined with large proofmass causes TFGs to have low resonant frequencies typically below 100 kHz. Consequently, they present low resistance to shock and random vibrations in the environment which can have frequencies up to 50 kHz. The poor vibration robustness does not necessarily affect their usage in low-end consumer applications such as gaming application, but it becomes problematic for high-end navigational applications. It has been shown that in the presence of random vibrations, the noise

Table 1.2 Summary of state-of-the-art performance of MEMS gyroscopes in commercial 3-axis gyroscope units or IMUs

Manufacturer	STM	InvenSense	Bosch	ADI
Model	LSM6DSM	IAM-20380	BMG250	ADIS16497
Production year	2017	2016	2016	2017
ARW (°/√h)[a]	0.228	0.3	0.42	0.09
Bias instability (°/h)	–	–	10	0.8
Frequency (kHz)	–	27	–	5.5
Package size (mm^3)	2.5 × 3.0 × 0.83	3.0 × 3.0 × 0.75	2.5 × 3.0 × 0.83	47 × 44 × 14
Unit price (USD)	4.09	–	4.19	2339.00

[a]From noise density

Table 1.3 Summary of state-of-the-art performance of 3-axis MEMS gyroscopes reported in academia

Manufacturer	UC Davis	Georgia Tech	UC Irvine
Year of publication	2015	2016	2017
ARW (°/√h)	2.16	21.42	7.2
Bias instability (°/h)	154.8	813.6	140.4
Frequency (kHz)	28	138	67
Die area (mm^2)	3.2 × 3.2	2.0 × 2.0	1.5 × 0.8

performance and stability of a low-frequency TFG can degrade by several orders of magnitude, making it unsuitable for navigational applications even with promising advertised performance. Tables 1.2 and 1.3 list the gyroscope performance and specifications of state-of-the-art commercial MEMS 3-axis gyroscopes or IMUs [35–38] and the performance of 3-axis MEMS gyroscopes recently reported in academia [39–41].

An alternative MEMS gyroscope technology is bulk acoustic wave (BAW) gyroscope operating in high-stiffness BAW modes with high resonant frequency in the megahertz range [42–44]. A silicon BAW disk yaw gyroscope was first reported by Houri Johari at Georgia Tech in 2006 [43]. The BAW disk gyroscope uses the degenerate $n = 3$ wine-glass modes at 6 MHz as operational modes. The high aspect-ratio poly- and single-crystal silicon (HARPSS™) process was used to fabricate BAW gyroscope, which enables nanoscale capacitive gaps to provide efficient transductions for the high-stiffness devices [45, 46]. The solid structure of BAW disk gyroscopes allows them to have superior robustness against shock and linear vibration. Recent advancement in BAW disk gyroscopes shows that consistent tactical-grade performance can be achieved in spite of the existence of random vibrations in the testing environment [47]. This robust high performance would allow BAW disk gyroscopes to provide reliable short-term (up to a few minutes) navigational guidance in complicated environment. In the case where intermittent external calibration like GPS or radio reference is available, an inertial-based reference-aided navigation system can also be realized (Fig. 1.8).

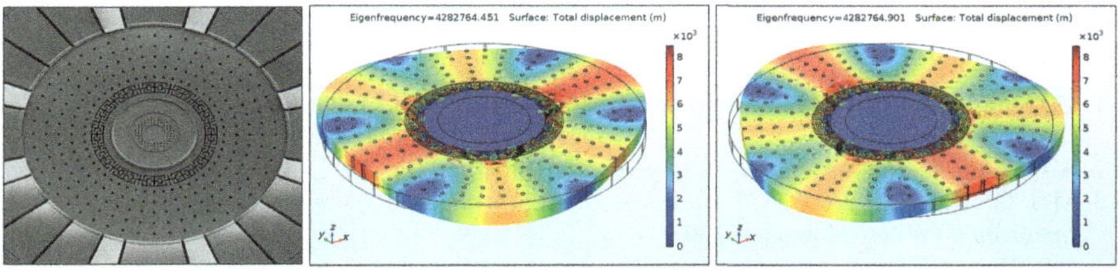

Fig. 1.8 Mechanical structure of a substrate-decoupled BAW disk gyroscope and mode-shapes of the degenerate $n = 3$ gyroscopic modes [47]

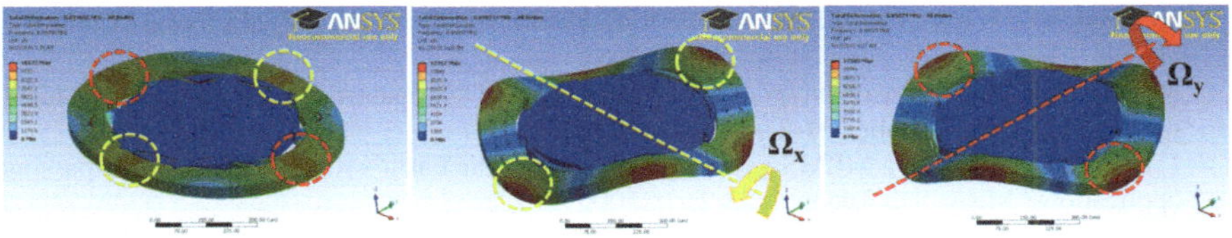

Fig. 1.9 Operational mode-shapes of the high-frequency annulus pitch and roll gyroscope [48]

Pitch and roll gyroscopes can also be implemented using stiff structures with high resonant frequencies. A high-frequency annulus gyroscope was reported in 2011 which is capable of sensing both pitch and roll rotations with an operational frequency of 0.9 MHz [48]. The annulus gyroscope uses the $n = 2$ IP wine-glass mode as the drive mode and a pair of orthogonal $n = 3$ OOP bending modes as sense modes; the theory of using such modes for gyroscope operation is explained in [49]. Like BAW disk yaw gyroscope, the high-frequency annulus is also expected to provide better robustness compared to their low-frequency counterparts. However, since pitch and roll gyroscopes involve non-degenerate operational modes with both IP and OOP degrees of freedom, they show more complicated dynamic behaviors and raise more design challenges. As a result, despite the inherent environmental robustness, compelling performance was not achieved with the annulus gyroscope (Fig. 1.9). Therefore, how to reach tactical and inertial-grade performance in pitch and roll gyroscopes with OOP degrees of freedom while preserving the robustness of high-frequency designs has become the biggest obstacle between MEMS technology and portable INS applications. This work aims to establish comprehensive understandings of the performance limits in high-frequency MEMS gyroscopes by studying physical phenomena in the gyroscope system. Based on the understandings, key factors are identified and addressed to significantly improve the gyroscope designs. Eventually, a single-chip MEMS IMU is developed, showing promising performance for navigational applications, and path toward further improvements is discussed.

References

1. G. Schmidt, INS/GPS technology trends. NATO RTO STO Educational Notes **116**(1), 1–24 (2011)
2. A.D. King, Inertial navigation-forty years of evolution. GEC Rev. **13**(3), 140–149 (1998)
3. M.S. Grewal, L.R. Weill, A.P. Andrews, *Global Positioning Systems, Inertial Navigation, and Integration* (Wiley, Hoboken, 2007)
4. E.B. Kaiser, E.B. Kaiser, M. Lawo. Wearable navigation system for the visually impaired and blind people, in *2012 IEEE/ACIS 11th International Conference on Computer and Information Science*, (2012), pp. 230–233
5. B. Barshan, H.F. Durrant-Whyte, Inertial navigation systems for mobile robots. IEEE Trans. Robot. Autom. **11**(3), 328–342 (1995)
6. H. Fourati et al., Posture and body acceleration tracking by inertial and magnetic sensing: application in behavioral analysis of free-ranging animals. Biomed. Signal Process. Control **6** (1), 94–104 (2011)
7. R. Li et al., A multisensor integration approach toward astronaut navigation for landed lunar missions. J. Field Robotics **31**(2), 245–262 (2014)
8. L.I. Iozan et al., Using a MEMS gyroscope to measure the Earth's rotation for gyrocompassing applications. Meas. Sci. Technol. **23**(2), 025005 (2012)
9. D.H. Eckhardt, Air Force Geophysics Lab Hanscom AFB MA, *Surveying and Geophysical Measurements with Inertial Rotation Sensors* (Defense Technical Information Center, Ft. Belvoir, 1978)
10. G.A. Sanders, M.G. Prentiss, S. Ezekiel, Passive ring resonator method for sensitive inertial rotation measurements in geophysics and relativity. Opt. Lett. **6**(11), 569–571 (1981)
11. O.J. Woodman, *An Introduction to Inertial Navigation* (University of Cambridge, Computer Laboratory, Cambridge, UK, 2007)
12. V. Marotto, et al. Orientation Analysis through a Gyroscope Sensor for Indoor Navigation Systems, in *Proceedings of 4th International Conference on Sensor Device Technology Applications*, (2013), pp. 85–90
13. E. Gorczycki, Dynamic considerations relating to the behavior of inertial space-stabilized platforms. J. Aeronaut. Sci. **24**(2), 130–138 (1957)
14. L. Ojeda, J. Borenstein. Personal dead-reckoning system for GPS-denied environments, in *2007 IEEE International Workshop on Safety, Security and Rescue Robotics*, (2007), pp. 1–6
15. C. Randell, C. Djiallis, H. Muller. Personal position measurement using dead reckoning, in *Proceedings of the 7th IEEE International Symposium on Wearable Computers* (IEEE Computer Society, 2003), pp. 166
16. D. Titterton, J.L. Weston, *Strapdown inertial navigation technology*, vol 17 (IET, Stevenage, UK, 2004)
17. F. Ayazi. Multi-DOF inertial MEMS: from gaming to dead reckoning, in *2011 16th International Solid-State Sensors, Actuators and Microsystems Conference*, (2011), pp. 2805–2808
18. N. Yazdi, F. Ayazi, K. Najafi, Micromachined inertial sensors. Proc. IEEE **86**(8), 1640–1659 (1998)
19. IMU of lunar modules used in the Apollo missions [Online], Available: http://www.hq.nasa.gov/alsj/lm_imu.gif
20. R.E. Mandapat, *Development and Evaluation of Positioning Systems for Autonomous Vehicle Navigation* (University of Florida, Gainesville, FL, USA, 2001)
21. A.I.U.l.e. Ishlinskiĭ, Mechanics of Gyroscopic Systems: (Mekhanika giroskopicheskikh sistem) (Israel Program for Scientific Translations, 1965); [available from the US Dept. of Commerce, Clearinghouse for Federal Scientific and Tecnical Information, Springfield, VA]
22. B.A. Shamir, *An Overview of Optical Gyroscopes Theory, Practical Aspects, Applications and Future Trends* [Online], Available: https://pdfs.semanticscholar.org/3fdb/a8cea7cf6d0cad17b13828970f763e368f43.pdf

23. This rare photo shows the most sensitive part of a combat aircraft [Online], Available: https://foxtrotalpha.jalopnik.com/this-rare-photo-shows-the-most-sensitive-part-of-a-comb-1627339307
24. KVH reveals plans to build FOG sensor for self-driving cars [Online], Available: https://insidegnss.com/kvh-reveals-plans-to-build-fog-sensor-for-self-driving-cars/
25. J. Fang, J. Qin, Advances in atomic gyroscopes: a view from inertial navigation applications. Sensors **12**(5), 6331 (2012)
26. Photo release – Northrop Grumman demonstrates micro-gyro prototype for DARPA program [Online], Available: https://news.northropgrumman.com/news/releases/photo-release-northrop-grumman-demonstrates-micro-gyro-prototype-for-darpa-program
27. V. Apostolyuk, *Coriolis Vibratory Gyroscopes* (Springer, Cham, 2016)
28. D.M. Rozelle. The hemispherical resonator gyro: from wineglass to the planets, in *Proceedings of 19th AAS/AIAA Space Flight Mechanics Meeting*, (2009), pp. 1157–1178
29. G.T.A. Kovacs, K. Petersen, M. Albin, Peer reviewed: silicon micromachining: sensors to systems. Anal. Chem. **68**(13), 407A–412A (1996)
30. P. Greiff, et al. Silicon monolithic micromechanical gyroscope, in *Solid-State Sensors and Actuators, 1991. Digest of Technical Papers, TRANSDUCERS '91, 1991 International Conference on*, (1991), pp. 966–968
31. W.A. Clark, R.T. Howe, R. Horowitz. Surface micromachined Z-axis vibratory rate gyroscope, in *Technical Digest Solid-State Sensor and Actuator Workshop*, (1996), pp. 283–287
32. J.A. Geen, et al. Single-chip surface-micromachined integrated gyroscope with 50/deg/hour root Allan variance, in *2002 IEEE International Solid-State Circuits Conference*. Digest of Technical Papers (Cat. No.02CH37315), vol. 1, 2002, pp. 426–427
33. J. Seeger, M. Lim, S. Nasiri. Development of high-performance, high-volume consumer MEMS gyroscopes, in *Solid-State Sensors, Actuators, and Microsystems Workshop*, (2010), pp. 61–64
34. Comparing the InvenSense and Bosch accelerometers found in the iPhone 6 [Online], Available: https://www.chipworks.com/about-chipworks/overview/blog/comparing-invensense-and-bosch-accelerometers-found-iphone-6
35. STM LSM6DSM, iNEMO inertial module: always-on 3D accelerometer and 3D gyroscope [Online], Available: http://www.st.com/en/mems-and-sensors/lsm6dsm.html
36. InvenSense-IAM-20380 high performance gyroscope [Online], Available: https://www.invensense.com/products/motion-tracking/3-axis/iam-20380/
37. BOSCH BMG250 low noise, low power triaxial gyroscope [Online], Available: https://www.bosch-sensortec.com/bst/products/all_products/bmg250
38. ADIS16497 tactical grade, six degrees of freedom inertial sensor [Online], Available: http://www.analog.com/en/products/mems/inertial-measurement-units/adis16497.html
39. S. Sonmezoglu et al., Single-structure micromachined three-axis gyroscope with reduced drive-force coupling. IEEE Electron Device Lett **36**(9), 953–956 (2015)
40. S. Wisher, et al. A high-frequency epitaxially encapsulated single-drive quad-mass tri-axial resonant tuning fork gyroscope, in *2016 IEEE 29th International Conference on Micro Electro Mechanical Systems (MEMS)*, (2016), pp. 930–933
41. A. Efimovskaya, et al. Compact roll-pitch-yaw gyroscope implemented in wafer-level epitaxial silicon encapsulation process, in *2017 IEEE International Symposium on Inertial Sensors and Systems (INERTIAL)*, (2017), pp. 1–2
42. J. Houri, A. Farrokh. High-frequency capacitive disk gyroscopes in (100) and (111) silicon, in *2007 IEEE 20th International Conference on Micro Electro Mechanical Systems (MEMS)*, (2007), pp. 47–50
43. H. Johari, F. Ayazi. Capacitive bulk acoustic wave silicon disk gyroscopes, in *2006 International Electron Devices Meeting*, (2006), pp. 1–4
44. A. Rahafrooz, et al. A 0.5 mm^2 7-MHz capacitive bulk acoustic wave gyroscope in (100) silicon with large dynamic range, in *2017 IEEE 30th International Conference on Micro Electro Mechanical Systems (MEMS)*, (2017), pp. 25–28

45. F. Ayazi, K. Najafi, High aspect-ratio combined poly and single-crystal silicon (HARPSS) MEMS technology. J. Microelectromech. Syst. **9**(3), 288–294 (2000)
46. S. Pourkamali, F. Ayazi. SOI-based HF and VHF single-crystal silicon resonators with SUB-100 nanometer vertical capacitive gaps, in TRANSDUCERS, Solid-State Sensors, Actuators and Microsystems, 12th International Conference on, 2003, vol. 1, (2003), pp. 837–840
47. D.E. Serrano et al., Substrate-decoupled, bulk-acoustic wave gyroscopes: design and evaluation of next-generation environmentally robust devices. Microsyst. Nanoeng. **2**, 16015 (2016)
48. W.K. Sung, M. Dalal, F. Ayazi. A mode-matched 0.9 MHz single proof-mass dual-axis gyroscope, in *2011 16th International Solid-State Sensors, Actuators and Microsystems Conference*, (2011), pp. 2821–2824
49. B.J. Gallacher, J.S. Burdess, A.J. Harris, Principles of a three-axis vibrating gyroscope. IEEE Trans. Aerosp. Electron. Syst. **37**(4), 1333–1343 (2001)

Chapter 2
The Physics of Resonant MEMS Gyroscopes

2.1 Principle of Operation

MEMS gyroscope belongs to the class of Coriolis vibratory gyroscope. CVG rely on the Coriolis effect, which was reported in 1835 by French scientist Gaspard-Gustave de Coriolis to describe the phenomenon of an inertial force acting on an object that is in motion relative to a rotating reference frame [1].

As shown in Fig. 2.1, when an object with mass m and a velocity of $\vec{v}_I$ with respect to an inertial frame is observed in a rotating frame with a rotation rate of $\vec{\Omega}$, the following identity can be obtained:

$$\vec{v}_I = \vec{v}_R + \vec{\Omega} \times \vec{r} \rightarrow \left(\frac{d\vec{r}}{dt}\right)_I = \left(\frac{d\vec{r}}{dt}\right)_R + \vec{\Omega} \times \vec{r}, \tag{2.1}$$

where the subscripts I and R indicate vectors expressed in the inertial frame and the rotating frame, respectively. Since all vectors act the same under coordinate transformation, replacing $\vec{r}$ by $\vec{v}$ in Eq. (2.1) gives:

$$\begin{aligned}\vec{a}_I &= \left(\frac{d^2\vec{r}}{dt^2}\right)_I = \left[\left(\frac{d}{dt}\right)_R + \vec{\Omega}\times\right]^2 \vec{r} \\ &= \left(\frac{d^2\vec{r}}{dt^2}\right)_R + 2\vec{\Omega} \times \left(\frac{d\vec{r}}{dt}\right)_R + \vec{\Omega} \times (\vec{\Omega} \times \vec{r}) \\ &= \vec{a}_R + 2\vec{\Omega} \times \vec{v}_R - \Omega^2\vec{r}\end{aligned} \tag{2.2}$$

H. Wen, *Toward Inertial-Navigation-on-Chip*, Springer Theses,
https://doi.org/10.1007/978-3-030-25470-4_2

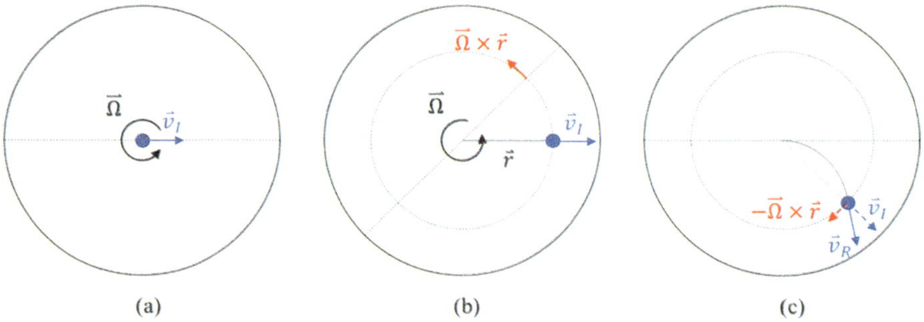

Fig. 2.1 Illustration of the Coriolis effect with (**a**) showing the initial state at time 0, (**b**) showing the observation in an inertial frame at time t, and (**c**) showing the observation in the rotating frame at time t

$$\vec{a}_R = \left(\vec{F}_I - 2m\vec{\Omega} \times \vec{v}_R + m\Omega^2\vec{r}\right)/m. \tag{2.3}$$

In Eq. (2.3), the left-hand side is the acceleration observed in the rotating frame. And the terms on the right-hand side represent the actual force applied, the Coriolis force, and the centrifugal force, respectively.

In 1851, French physicist Léon Foucault performed the Foucault pendulum experiment that demonstrated the Coriolis effect where a pendulum appears to experience a force orthogonal to the plane of swing due to the earth's rotation [2]. Similar operation is adapted to enable MEMS vibratory gyroscopes. In a MEMS gyroscope, the pendulum is replaced by a microscale spring-mass element, and instead of swinging under the action of gravity, the element is driven by an external alternating force to oscillate in a resonant mode, which is called the drive mode. Under the action of rotations orthogonal to the drive-mode motion, the element will experience an alternating Coriolis force with the same frequency as the drive-mode vibration as given in (2.3). The Coriolis force then excites the element to move in a secondary mode, i.e., the sense mode, where the displacement amplitude of the sense mode is proportional to the Coriolis force and therefore the rotation rate. The ratio between the sense-mode displacement amplitude and the rotation rate is called the mechanical scale-factor of the gyroscope (Fig. 2.2).

For example, in a yaw (z-axis) MEMS TFG [3], a pair of proofmass is connected and suspended through flexible beams. The proofmass is actuated to vibrate along x-axis. According to (2.3), when the gyroscope is rotated around z-axis, the Coriolis force causes the proofmass to vibrate along the y-direction at the drive frequency with an amplitude proportional to the rotation rate (Fig. 2.3).

The dynamics of a gyroscope is described by the equations of motion of the drive and sense modes, which take the form of two orthogonal second-order differential equations coupled by Coriolis force terms. Ignoring the centrifugal force terms, the matrix form of the equations of motion for an ideal MEMS gyroscope is given by:

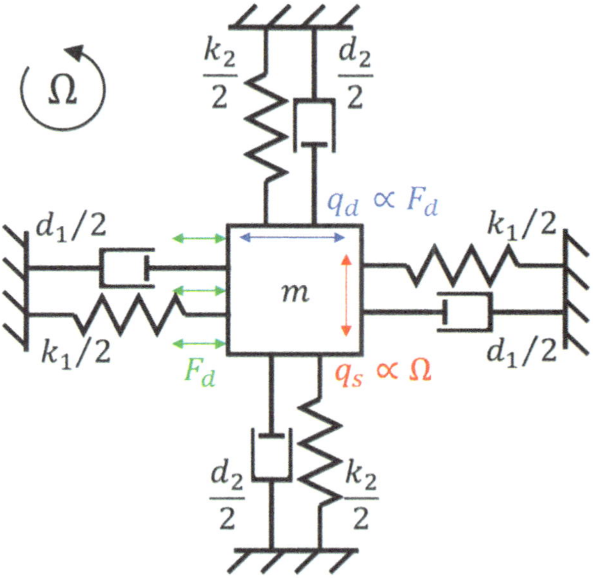

Fig. 2.2 Lumped-element illustration of the CVG operational principle

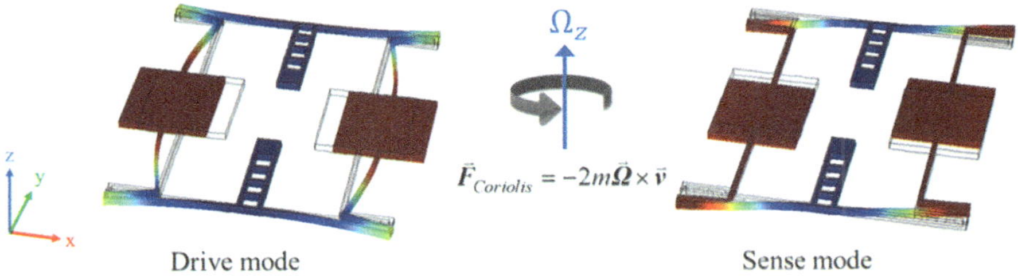

Fig. 2.3 Operational mode shapes of a yaw TFG

$$m\begin{bmatrix} \ddot{q}_1 \\ \ddot{q}_2 \end{bmatrix} + \begin{bmatrix} d_1 & 2\lambda m\Omega \\ -2\lambda m\Omega & d_2 \end{bmatrix}\begin{bmatrix} \dot{q}_1 \\ \dot{q}_2 \end{bmatrix} + \begin{bmatrix} k_1 & 0 \\ 0 & k_2 \end{bmatrix}\begin{bmatrix} q_1 \\ q_2 \end{bmatrix} = \begin{bmatrix} F_{\mathrm{d}} \\ 0 \end{bmatrix}, \tag{2.4}$$

where d_1, k_1, and d_2, k_2 are the damping coefficient and spring constant of the drive mode and sense mode, respectively. F_{d} is the driving force, and q_1, q_2 represent the displacement of each mode at the locations with maximum vibration amplitude, i.e., the antinodes. Mode shape-dependent constants m and λ are the effective mass of the operational modes and the Coriolis-coupling coefficient given by:

$$m = \rho \oint \left|\vec{U}_1\right|^2 dV = \rho \oint \left|\vec{U}_2\right|^2 dV, \text{ and } \lambda = \frac{\rho \oint \left|\vec{U}_1 \times \vec{U}_2 \cdot \hat{\Omega}\right| dV}{m}, \tag{2.5}$$

where $\vec{U}_1$ and $\vec{U}_2$ are normalized mode-shape function for the drive mode and sense mode describing normalized displacement field at different locations. The eigenfrequency, quality-factor (Q-factor), and time constant of each mode are:

$$\omega_i = \sqrt{\frac{k_i}{m}}, Q_i = \frac{m\omega_i}{d_i}, \text{ and } \tau_i = \frac{2Q_i}{\omega_i}. \tag{2.6}$$

Substitute (2.6) into (2.4), for a sinusoidal driving force with angular frequency ω_0, the equations of motion can be written as:

$$\left(-\omega_0^2 + j\omega_0\begin{bmatrix} \omega_1/Q_1 & 2\lambda\Omega \\ -2\lambda\Omega & \omega_2/Q_2 \end{bmatrix} + \begin{bmatrix} \omega_1^2 & 0 \\ 0 & \omega_2^2 \end{bmatrix}\right)\begin{bmatrix} q_1 \\ q_2 \end{bmatrix} = \begin{bmatrix} F_{\mathrm{d}}/m \\ 0 \end{bmatrix}. \tag{2.7}$$

Assuming the Coriolis force caused by sense motion is much smaller than the driving force, q_1 and q_2 are solved to be:

$$q_1 = \frac{F_{\mathrm{d}}}{k_1}\frac{1}{1 - \left(\frac{\omega_0}{\omega_1}\right)^2 + j\frac{1}{Q_1}\left(\frac{\omega_0}{\omega_1}\right)} \tag{2.8}$$

$$q_2 = j\frac{2\lambda q_1\Omega\omega_0}{\omega_2^2}\frac{1}{1 - \left(\frac{\omega_0}{\omega_2}\right)^2 + j\frac{1}{Q_2}\left(\frac{\omega_0}{\omega_2}\right)}. \tag{2.9}$$

For a given driving force amplitude, q_1 has the largest amplitude when $\omega_0 = \omega_1$. Substituting $\omega_0 = \omega_1$ into (2.8) gives:

$$\left|\frac{q_1}{F_d}\right| = \frac{Q_1}{k_1}, \text{ and } \angle\frac{q_1}{F_d} = -90^\circ. \tag{2.10}$$

Equation (2.10) shows that Q_1 times higher drive amplitude can be achieved when the gyroscope is actuated at its drive-mode natural frequency compared to a quasi-static drive-mode response with an off-resonance actuation force. For well-designed MEMS resonators, the quality-factor can range from a few hundreds to a few millions. Therefore, in practice it is always favorable to operate the gyroscope with $\omega_0 = \omega_1$ to achieve optimum actuation efficiency and minimize the power consumption.

The mechanical scale-factor of the gyroscope with a fixed drive amplitude can be found from (2.9):

$$\mathrm{SF_{mech}} = \left|\frac{q_2}{\Omega}\right| = \frac{\omega_0}{\omega_2^2}\frac{2\lambda|q_1|}{\sqrt{\left(1-\frac{\omega_0^2}{\omega_2^2}\right)^2+\left(\frac{1}{Q_2}\frac{\omega_0}{\omega_2}\right)^2}}. \tag{2.11}$$

Depending on the relation between the drive frequency and the sense-mode natural frequency, two situations are of interest:

$$\mathrm{SF_{mech}}|_{\omega_0 \ll \omega_2} \approx 2\lambda|q_1|\frac{\omega_0}{\omega_2^2} \text{ with } \angle\frac{q_2}{q_1} = -90^{\circ} \tag{2.12}$$

$$\mathrm{SF_{mech}}|_{\omega_0 = \omega_2} = 2\lambda|q_1|\frac{Q_2}{\omega_0} \text{ with } \angle\frac{q_2}{q_1} = 0^{\circ}. \tag{2.13}$$

The situation represented by (2.12) is called mode-split (or quasi-static) operation. The condition of $\omega_0 \ll \omega_2$ is easy to achieve by design and is insensitive to small structure variations from fabrication errors, therefore mode-split operation is widely used in low-end consumer MEMS TFGs to bring fast profits with minimum design investment and fabrication cost. However, since the scale-factor of mode-split gyroscopes is inversely proportional to the square of sense-mode resonant frequency, low-stiffness designs must be used to obtain reasonable performance. As pointed out in Chap. 1, the low-stiffness and low-frequency designs show poor robustness and cannot be used to fulfill high-end tasks in complicated environment for emerging navigational applications.

On the other hand, (2.13) describes the mode-matched (or resonant) operation of a MEMS gyroscope. In mode-matched operation, the sense mode is excited by the Coriolis force at its resonant frequency, which leads to a mechanical Q-amplification in the sense response. With the sense-mode Q ranging from hundreds to millions, orders of magnitude higher scale-factor can be achieved compared to the mode-split operation even with designs having higher stiffness and higher resonant frequencies. This allows resonant gyroscopes to use robust structures that are capable of delivering consistent high performance in practical usage (Fig. 2.4).

Studying the fundamental operation of MEMS gyroscopes provides the general guideline that high-frequency high-Q resonant operation is essential for achieving

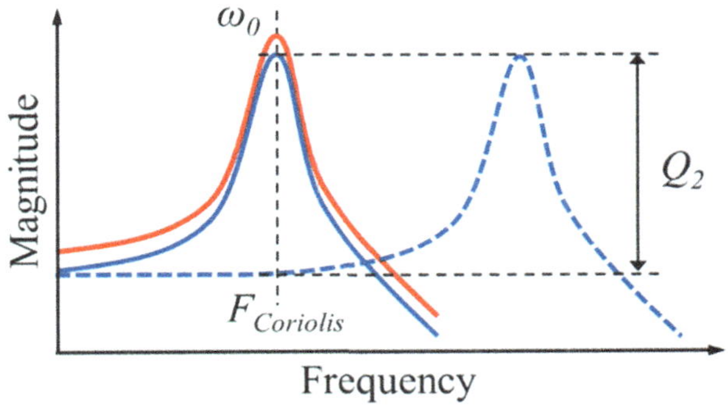

Fig. 2.4 Comparison between the sense response of mode-matched (solid-blue line) and mode-split (dashed-blue line) operation

robust high-performance gyroscopes. In practice, imperfections in materials, fabrication, and interface electronics will introduce various errors that may degrade the gyroscope performance. Additional design considerations must be given to address different issues and improve the limiting factors.

2.2 Non-ideal MEMS Gyroscope

2.2.1 Mechanical Model

In general, a MEMS gyroscope using nondegenerate operational modes will have different effective mass, effective stiffness, and damping mechanism for the drive mode and sense mode. Modifications must be made to (2.4) to capture the difference and predict accurate dynamics of the gyroscope. The effects of imperfections on gyroscope equations of motion were discussed in [4] with a point-mass lumped model. Here, we further generalize the model by taking into account the mode-shape difference for nondegenerate gyroscopes like annulus pitch or roll gyroscopes using both IP and OOP high-frequency modes. For gyroscopes using resonant modes with unalike mode shapes, the lumped model of the unforced loss-less vibration along the two orthogonal principle resonance axes is:

$$\begin{bmatrix} m_1 & 0 \\ 0 & m_2 \end{bmatrix} \begin{bmatrix} \ddot{q}_1 \\ \ddot{q}_2 \end{bmatrix} + \begin{bmatrix} 0 & 2\gamma\Omega \\ -2\gamma\Omega & 0 \end{bmatrix} \begin{bmatrix} \dot{q}_1 \\ \dot{q}_2 \end{bmatrix} + \begin{bmatrix} k_{11} & 0 \\ 0 & k_{22} \end{bmatrix} \begin{bmatrix} q_1 \\ q_2 \end{bmatrix} = \begin{bmatrix} 0 \\ 0 \end{bmatrix}, \quad (2.14)$$

where the effective mass and Coriolis-coupled mass are defined as:

$$m_1 = \rho \oint \left| \vec{U}_1 \right|^2 dV, m_2 = \rho \oint \left| \vec{U}_2 \right|^2 dV, \text{and } \gamma = \rho \oint \left| \vec{U}_1 \times \vec{U}_2 \cdot \hat{\Omega} \right| dV. \quad (2.15)$$

Now, considering material and fabrication imperfections such as wafer crystal orientation misalignment and wafer thickness variations, the principle resonance axis of the operational modes may not be aligned with the forcer and pickoff electrodes. Therefore, a coordinate transformation needs to be applied to (2.14) to evaluate the observed response along the detection axes. Similarly, process errors like lithography inaccuracy and electrode gap-size variations may alter the energy dissipation path and cause a misalignment between the principle damping axes and actuation/readout axes. The complete model including damping and actuation terms is given by:

$$
\begin{aligned}
&\begin{bmatrix} m_1 & 0 \\ 0 & m_2 \end{bmatrix}_\omega \begin{bmatrix} \ddot{q}_{\omega 1} \\ \ddot{q}_{\omega 2} \end{bmatrix} + \begin{bmatrix} 0 & 2\gamma\Omega \\ -2\gamma\Omega & 0 \end{bmatrix}_\omega \begin{bmatrix} \dot{q}_{\omega 1} \\ \dot{q}_{\omega 2} \end{bmatrix} + \begin{bmatrix} k_{11} & 0 \\ 0 & k_{22} \end{bmatrix}_\omega \begin{bmatrix} q_{\omega 1} \\ q_{\omega 2} \end{bmatrix} \\
&\quad + \begin{bmatrix} d_1 & 0 \\ 0 & d_2 \end{bmatrix}_\tau \begin{bmatrix} \dot{q}_{\tau 1} \\ \dot{q}_{\tau 2} \end{bmatrix} \\
&\quad = \begin{bmatrix} F_d \\ 0 \end{bmatrix},
\end{aligned}
\tag{2.16}
$$

where notations ω and τ are used to indicate the evaluation coordinate system (Fig. 2.5). Consider the following rotation matrices:

$$
\boldsymbol{R}_\omega = \begin{bmatrix} \cos\theta_\omega & -\sin\theta_\omega \\ \sin\theta_\omega & \cos\theta_\omega \end{bmatrix}, \text{and } \boldsymbol{R}_\tau = \begin{bmatrix} \cos\theta_\tau & -\sin\theta_\tau \\ \sin\theta_\tau & \cos\theta_\tau \end{bmatrix}. \tag{2.17}
$$

The equations of motion can be expressed in the observation coordinate bases through coordinate transformation:

$$
\begin{aligned}
&\boldsymbol{R}_\omega \boldsymbol{M}_\omega \boldsymbol{R}_\omega^{-1} \begin{bmatrix} \ddot{q}_1 \\ \ddot{q}_2 \end{bmatrix} + \boldsymbol{R}_\omega \boldsymbol{K}_\omega \boldsymbol{R}_\omega^{-1} \begin{bmatrix} q_1 \\ q_2 \end{bmatrix} + \boldsymbol{R}_\tau \boldsymbol{D}_\tau \boldsymbol{R}_\tau^{-1} \begin{bmatrix} \dot{q}_1 \\ \dot{q}_2 \end{bmatrix} \\
&\quad + 2\gamma\Omega \begin{bmatrix} 0 & 1 \\ -1 & 0 \end{bmatrix} \begin{bmatrix} \dot{q}_1 \\ \dot{q}_2 \end{bmatrix} \\
&\quad = \begin{bmatrix} F_d \\ 0 \end{bmatrix}
\end{aligned}
\tag{2.18}
$$

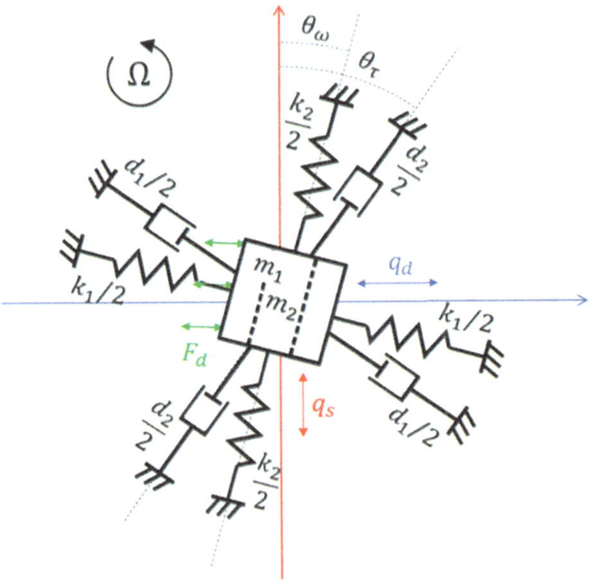

Fig. 2.5 Lumped model of a non-ideal nondegenerate gyroscope

$$\rightarrow \boldsymbol{M}\begin{bmatrix}\ddot{q}_1\\ \ddot{q}_2\end{bmatrix}+\left(\boldsymbol{D}+2\gamma\Omega\begin{bmatrix}0 & 1\\ -1 & 0\end{bmatrix}\right)\begin{bmatrix}\dot{q}_1\\ \dot{q}_2\end{bmatrix}+\boldsymbol{K}\begin{bmatrix}q_1\\ q_2\end{bmatrix}=\begin{bmatrix}F_d\\ 0\end{bmatrix}, \tag{2.19}$$

where the elements of mass, stiffness, and damping matrices after coordinate transformations are:

$$m_{11}=m+\Delta m\cos 2\theta_\omega, m_{22}=m-\Delta m\cos 2\theta_\omega, m_{12}=\Delta m\sin 2\theta_\omega \tag{2.20}$$

$$k_{11}=k+\Delta k\cos 2\theta_\omega, k_{22}=k-\Delta k\cos 2\theta_\omega, k_{12}=\Delta k\sin 2\theta_\omega \tag{2.21}$$

$$d_{11}=d+\Delta d\cos 2\theta_\tau, d_{22}=d-\Delta d\cos 2\theta_\tau, d_{12}=\Delta d\sin 2\theta_\tau \tag{2.22}$$

$$\text{with } c=\frac{c_1+c_2}{2} \text{ and } \Delta c=\frac{c_1-c_2}{2}, \text{ where } c\in\{m,k,d\}. \tag{2.23}$$

Equation (2.19) gives the generalized gyroscope equations of motion, which can be used to study the behaviors of any vibratory rotation-rate gyroscopes with open-loop-sensing operation. For closed-loop operation, the equations can easily be adapted by including the restoring force term along the sensing direction.

2.2.1.1 Mode-Matching

The undamped eigenfrequencies of the non-ideal gyroscope are obtained by solving (2.19) ignoring the damping, Coriolis force, and actuations force terms:

$$\omega_n^2=\left(\frac{\omega_{11}^2+\omega_{22}^2}{2}-\frac{k_{12}m_{12}}{m_{11}m_{22}}\pm\frac{\Delta\omega^2}{2}\right)/\left(1-\frac{m_{12}^2}{m_{11}m_{22}}\right) \tag{2.24}$$

$$\begin{aligned}\omega_{11}^2&=\frac{k_{11}}{m_{11}},\ \omega_{22}^2=\frac{k_{22}}{m_{22}}\\ \Delta\omega^2&=\sqrt{\left(\omega_{11}^2-\omega_{22}^2\right)^2+\frac{4\left(k_{12}-m_{12}\omega_{11}^2\right)\left(k_{12}-m_{12}\omega_{22}^2\right)}{m_{11}m_{22}}}.\end{aligned} \tag{2.25}$$

Solve for $\Delta\omega=0$, and we get:

$$k_{12}=\frac{m_{12}\left(\omega_{11}^2+\omega_{22}^2\right)}{2}\pm\frac{1}{2}\sqrt{-\left(m_{11}m_{22}-m_{12}^2\right)\left(\omega_{11}^2-\omega_{22}^2\right)^2}. \tag{2.26}$$

From (2.20) and (2.23), we have:

$$m_{11}m_{22} - m_{12}^2 = m^2 - \Delta m^2 = m_1 m_2 > 0. \tag{2.27}$$

Therefore, for real values of k_{12}, the frequency split between the two eigenmodes is zero (i.e., mode-matched) only when:

$$\frac{k_{11}}{m_{11}} = \frac{k_{22}}{m_{22}}, \text{and } k_{12} = m_{12}\frac{k_{11}}{m_{11}} = m_{12}\frac{k_{22}}{m_{22}}. \tag{2.28}$$

Substituting (2.28) into (2.24) gives the following matched frequency:

$$\omega_1 = \omega_2 = \sqrt{\frac{k_{11}}{m_{11}}} = \sqrt{\frac{k_{22}}{m_{22}}}. \tag{2.29}$$

2.2.1.2 Sense Output Bias

Define the Coriolis-coupling coefficient to be $\lambda = \gamma/m_{22}$; solving (2.19) for the sense-mode displacement gives:

$$q_2 = B_{\text{disp}} + q_{2\text{Coriolis}}(\Omega) + O(\Omega^2) \tag{2.30}$$

$$q_{2\text{Coriolis}}(\Omega) = \frac{2j\lambda m_{22}\omega_0 q_1}{k_{22} - \omega_0^2 m_{22} + j\omega_0 d_{22}}\Omega \tag{2.31}$$

$$B_{\text{disp}} = -q_1 \frac{k_{12} - \omega_0^2 m_{12} + j\omega_0 d_{12}}{k_{22} - \omega_0^2 m_{22} + j\omega_0 d_{22}}. \tag{2.32}$$

Expression (2.31) gives the rotation-induced sense-mode displacement of the non-ideal gyroscope, and (2.32) gives the sense-mode displacement bias due to cross-mode coupling through stiffness and damping axes misalignment. The imaginary part of the numerator in B_{disp} causes bias that is in phase with the Coriolis response, and the real part of the numerator causes bias that is in quadrature phase with the Coriolis response. When mode-matching conditions (2.28) are met, (2.31) recovers to the form of an ideal mode-matched gyroscope:

$$|q_{2\text{Coriolis}}| = 2\lambda|q_1|\frac{m_{22}}{d_{22}}\Omega = 2\lambda|q_1|\frac{Q_2}{\omega_0}\Omega \text{ and } \angle\frac{q_2}{q_1} = 0^\circ, \tag{2.33}$$

with the effective sense-mode quality-factor given by the transformed damping factor d_{22} along the detection axis. The mechanical scale-factor is:

$$\mathrm{SF}_{\mathrm{mech}} = \left|\frac{q_{2\mathrm{Coriolis}}}{\Omega}\right| = 2\lambda|q_1|\frac{Q_2}{\omega_0}. \tag{2.34}$$

Combining (2.31) and (2.32) gives the bias equivalent rotation rate output:

$$B_\Omega = \frac{-\omega_0 d_{12} + j\left(k_{12} - \omega_0^2 m_{12}\right)}{2\lambda m_{22}\omega_0}. \tag{2.35}$$

The first term in the numerator represents the bias component that is in phase with Coriolis-induced signal. The in-phase bias is an output offset that is indistinguishable from real rotation rate response, and it is also referred to as zero-rate output (ZRO) sometime. The ZRO originates from the damping mismatch and the principal damping axis misalignment. In general, the ZRO should be kept small and stable by design, especially for pitch and roll gyroscopes using operational modes with different dissipation mechanisms, which will be discussed in detail in the following chapters. The second term in the numerator is the bias component that is in quadrature phase with rate response. Ideally, quadrature signal is rejected by in-phase demodulation of the sense signal. However, in practice, frequency and phase variations of the gyroscope system can cause significant demodulation errors if quadrature level is not minimized. More importantly, the cross-coupling stiffness causing quadrature also leads to frequency veering phenomenon and prevents the mode-matched operation of the gyroscope [5].

2.2.2 *Electromechanical Transduction and Interface Circuits*

The previous sections described the mechanical model of resonant gyroscopes. The complicated MEMS gyroscope systems also include electromechanical transductors and conditioning electronics to convert physical signals between mechanical domain and electrical domain and to provide control and compensation for mechanical imperfections. The most commonly used transduction mechanisms in integrated MEMS are capacitive transduction and piezoelectric transduction. Piezoelectric transductors generally provide higher transduction coupling efficiency. However, it has several major drawbacks when used for MEMS gyroscopes. First, silicon itself as the standard substrate for MEMS devices is not a piezoelectric material. To implement piezoelectric transductions in silicon MEMS, additional layers need to be deposited, which introduces large material interface loss and lowers the quality-factors of the gyroscope. This not only affects the drive-loop phase noise, sense thermal noise, etc., but it also makes the Q unpredictable for the gyroscope, risking the design reliability. Second, as will be shown later in this chapter, MEMS gyroscopes usually require tuning or compensation for fabrication non-idealities, which are not offered by piezoelectric transduction. In addition, piezoelectric transducers may also have other limits such as fabrication feasibility and mode-shape compatibility issues.

In comparison, capacitive transductions allow cleaner resonant response, enable electrostatic tuning, and have better integrability. The shortcoming of capacitive transduction is a lower coupling efficiency. This could be limiting the noise and power performance of a MEMS gyroscope, especially high-stiffness designs like BAW gyroscopes. However, we will show that using the HARPSS™ process, efficient transduction can be achieved with nanoscale capacitive gaps even for BAW high-frequency devices.

2.2.3 *Drive-Loop Transduction and Control*

In MEMS rotation rate gyroscopes, the drive mode of the gyroscope is excited and sustained in oscillation with a fixed amplitude at its resonant frequency. In parallel-plate capacitive transduction, the electrostatic force of a forcer electrode is:

$$F_d = \frac{1}{2}\frac{\partial C}{\partial q_1}(V_P + v_d)^2 \approx \frac{1}{2}\frac{\varepsilon A}{g_0^2}\left(V_P^2 + 2V_P v_d + v_d^2\right), \tag{2.36}$$

where A, g_0, and C are the transduction area, initial gap-size, and capacitance of the actuation capacitor. V_P is a DC polarization applied between the resonator and the electrode, and v_d is the AC actuation voltage. The approximate equality is valid for linear actuation range with drive amplitude less than one-tenth the gap-size. There are two AC force components, one at the frequency of v_d and one at double the frequency. The first AC term is scaled by the DC polarization and provides larger actuation force. Therefore, a DC polarization voltage and an AC voltage having the same frequency as the gyroscope resonant frequency are used for exciting the drive mode. The drive-mode motion can be detected with another capacitor electrode:

$$i_d = V_P\frac{\partial C}{\partial t} = V_P\frac{\partial C}{\partial q_1}\frac{\partial q_1}{\partial t} \approx \frac{\varepsilon A V_P}{g_0^2}\dot{q}_1. \tag{2.37}$$

Combining (2.10), (2.36), and (2.37) gives:

$$|i_d| = \left(\frac{\varepsilon A V_P}{g_0^2}\right)^2 \frac{Q_1}{\omega_1 m_1}|v_d|. \tag{2.38}$$

Depending on the electrode locations and mode-shape symmetry, i_d and v_d can either be in phase or 180° out of phase. From (2.38) we can also get the motional impedance of a gyroscope:

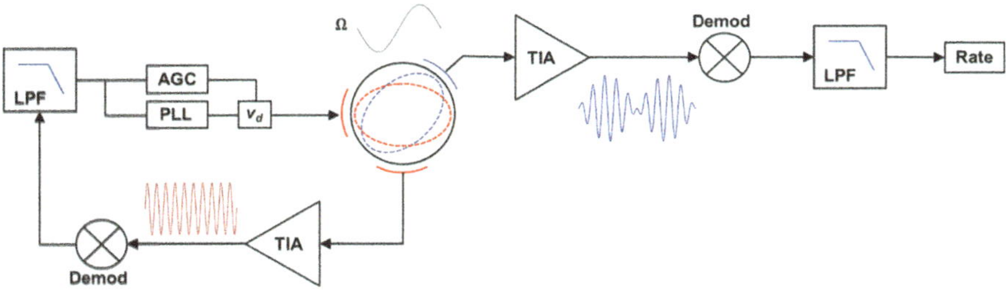

Fig. 2.6 Block diagram of a gyroscope system with drive and sense electronics. TIA stands for transimpedance amplifier

$$R_m = \left|\frac{v_d}{i_d}\right| = \frac{g_0^4}{(\varepsilon A V_P)^2}\frac{\omega_1 m_1}{Q_1}. \tag{2.39}$$

Equation (2.39) shows that the gyroscope drive-mode motional impedance is proportional to g_0^4. By going to nanoscale capacitive gaps instead of conventional 1 to 10 μm gaps, the motional impedance increases due to a higher resonant frequency can easily be overcome.

The drive oscillation is established by closing a phase-locked loop (PLL), which tracks the drive-mode resonant frequency with a fixed phase relation determined by the i_d-v_d relation and additional phase shift from electronics. Another automatic gain control (AGC) loop can also be added to reduce variations in actuation voltage and maintain fixed drive amplitude (Fig. 2.6).

2.2.4 *Sense-Mode Detection*

Similarly, the sense-mode motion can also be detected using capacitive electrodes with nanoscale gaps. From (2.30) and (2.37) we have:

$$i_s \approx \frac{\varepsilon A V_P}{g_0^2}\dot{q}_2 = i_q + i_i \tag{2.40}$$

$$i_q = j\frac{\varepsilon A V_P \omega_0 q_1}{g_0^2}\frac{\omega_0^2 m_{12} - k_{12}}{k_{22} - \omega_0^2 m_{22} + j\omega_0 d_{22}} \tag{2.41}$$

$$i_i = \frac{\varepsilon A V_P \omega_0 q_1}{g_0^2}\frac{\omega_0 d_{12} - 2\lambda m_{22}\omega_0 \Omega}{k_{22} - \omega_0^2 m_{22} + j\omega_0 d_{22}}, \tag{2.42}$$

where i_q and i_i are the quadrature and in-phase components of the sense channel output current, respectively. Through proper demodulation, the quadrature component is rejected, and the in-phase component is read out to determine the rotation rate. With mode-matched condition, (2.42) gives:

$$i_i = -j\frac{\varepsilon A V_P Q_2 q_1}{g_0^2}\frac{d_{12}}{m_{22}} + 2j\lambda Q_2 q_1 \Omega \frac{\varepsilon A V_P}{g_0^2}. \tag{2.43}$$

The first term represents the ZRO current, and the second term gives the mode-matched current scale-factor as:

$$\mathrm{SF_{elec}} = \left|\frac{i_s}{\Omega}\right| = 2\lambda Q_2 |q_1| \frac{\varepsilon A V_P}{g_0^2} \text{ and } \angle\frac{i_s}{q_1} = 90^\circ. \tag{2.44}$$

Here the phase relation is given based on the generalized displacement variable. In practice, depending on the mode-shape symmetry and sensing electrode location, the phase difference can also be $-90°$. Note the electrical current scale-factor is independent of the gyroscope frequency. However, by using small gap, the current scale-factor can be improved.

2.2.5 *Electrostatic Tuning*

Material and fabrication non-idealities will cause a MEMS gyroscope to present mismatched as-fabricated natural modes and principle resonant axes misaligned with the designated electrodes, which are shown as mismatched diagonal terms and non-zero off-diagonal terms in the transformed mechanical stiffness matrix. To recover mode-matched operation, the mechanical stiffness matrix needs to be modified to meet the conditions given by (2.28). This is done through electrostatic tuning [6]. For a parallel-plate capacitive transducer between the resonator and a fixed electrode, the electrical energy stored in the capacitor is:

$$U_e = \frac{1}{2}V_{\mathrm{DC}}^2 C(q_1, q_2) = \frac{V_{\mathrm{DC}}^2}{2}\frac{\varepsilon A}{g(q_1, q_2)}, \tag{2.45}$$

where V_{DC} is the DC voltage difference between the resonator and the electrode. Depending on the electrode location with respect to the operational mode shapes, the gap-size can be a function of either drive or sense-mode displacement or both. The displacement-dependent potential energy generates an electrostatic stiffness matrix given by:

$$\boldsymbol{K}_{\text{elec}} = \begin{bmatrix} k_{e11} & k_{e12} \\ k_{e12} & k_{e22} \end{bmatrix} = -\begin{bmatrix} \dfrac{\partial^2 U_e}{\partial q_1^2} & \dfrac{\partial^2 U_e}{\partial q_1 \partial q_2} \\ \dfrac{\partial^2 U_e}{\partial q_1 \partial q_2} & \dfrac{\partial^2 U_e}{\partial q_2^2} \end{bmatrix}$$
$$= -\frac{V_{\text{DC}}^2}{2}\begin{bmatrix} \dfrac{\partial^2 C}{\partial q_1^2} & \dfrac{\partial^2 C}{\partial q_1 \partial q_2} \\ \dfrac{\partial^2 C}{\partial q_1 \partial q_2} & \dfrac{\partial^2 C}{\partial q_2^2} \end{bmatrix}. \tag{2.46}$$

The equations of motion are then given by:

$$\boldsymbol{M}\begin{bmatrix} \ddot{q}_1 \\ \ddot{q}_2 \end{bmatrix} + \left(\boldsymbol{D} + 2\lambda m_2 \Omega \begin{bmatrix} 0 & 1 \\ -1 & 0 \end{bmatrix}\right)\begin{bmatrix} \dot{q}_1 \\ \dot{q}_2 \end{bmatrix}$$
$$+ \left(\boldsymbol{K}_{\text{mech}} + \sum_n \boldsymbol{K}_{\text{elec\_}n}\right)\begin{bmatrix} q_1 \\ q_2 \end{bmatrix}$$
$$= \begin{bmatrix} F_d \\ 0 \end{bmatrix}, \tag{2.47}$$

where $\boldsymbol{K}_{\text{elec}\ \_\ n}$ represents the electrostatic stiffness contribution from the nth electrode of the gyroscope. By changing the DC tuning voltages at different electrodes, the overall stiffness can be adjusted to meet the mode-matching criteria given in (2.28).

2.2.5.1 Stiffness Mismatch Tuning

For an electrode with capacitive gap perpendicular to the drive-mode detection axes in the observing coordinate, the capacitance as a function of mode displacement and the corresponding electrostatic stiffness are:

$$C = \frac{\varepsilon A}{(g_0 - q_1)} = \frac{\varepsilon A}{g_0}\left\{1 + \frac{q_1}{g_0} + \left(\frac{q_1}{g_0}\right)^2 + \ldots\right\} \text{ and } \boldsymbol{K}_{\text{elec}} = -\frac{\varepsilon A V_{DC}^2}{g_0^3}\begin{bmatrix} 1 & 0 \\ 0 & 0 \end{bmatrix}, \tag{2.48}$$

which provide only a non-zero diagonal element for direct frequency tuning of the drive mode. Similarly, an electrode with gap perpendicular to the sense-mode detection axes will produce an electrostatic stiffness matrix with only non-zero k_{e22}. The diagonal electrostatic spring constants are always negative despite the polarity of the voltage difference and displacement, which is called the electrostatic spring softening effect and can be used to tune down the frequency of the higher-frequency mode and reduce frequency split. When tuning condition $(k_{11} + k_{e11})/m_{11} = (k_{22} + k_{e22})/m_{22}$ is met, the gyroscope modes reach a minimum frequency split defined by (2.24) and (2.25) as:

$$\left|\omega_1^2 - \omega_2^2\right|_{\min} = \sqrt{\frac{4\left(k_{12} - m_{12}\omega_{11}^2\right)^2}{m_{11}m_{22}}} \Big/ \left(1 - \frac{m_{12}^2}{m_{11}m_{22}}\right). \tag{2.49}$$

For a drive and sense-mode-symmetric case ($m_1 = m_2 = m$), it gives:

$$\left|\omega_1^2 - \omega_2^2\right|_{min} = \left|\frac{2k_{12}}{m}\right|. \tag{2.50}$$

This minimum tunable frequency-split phenomenon is called frequency veering effect which is commonly seen in coupled harmonic oscillators. The existence of veering effect in non-ideal gyroscopes is expected by the non-zero k_{12} term appeared in the transformed equations of motion. However, a better physical interpretation of the phenomenon can be achieved by revisiting the problem in the gyroscope's principle-stiffness-axis coordinate. Substituting Eqs. (2.20), (2.21), (2.22) and (2.23) into the eigenfrequency expression (2.24) gives:

$$\omega_n^2 = \omega_{1,2}^2 = \frac{k_1}{m_1} = \frac{k_2}{m_2}. \tag{2.51}$$

This means the observed resonant frequencies follows the intrinsic resonant frequencies of the two modes, and mode-matched operation is achieved only when the two intrinsic resonances are matched. When a tuning voltage is applied along the observing axes, the electrostatic force it generates is misaligned from the principle stiffness axis. Therefore, it not only reduces mismatch between k_1 and k_2 but also rotates the principle stiffness axes with respect to the observing axes. Considering a mode-symmetric ($m_1 = m_2 = m$) example, a minimum observed frequency split is reached when:

$$k_{11} = k_{22} \text{ and } k_{12} \neq 0. \tag{2.52}$$

Substituting (2.21) into (2.52) gives:

$$k + \Delta k \cos 2\theta_\omega = k - \Delta k \cos 2\theta_\omega \text{ and } \Delta k \sin 2\theta_\omega \neq 0 \rightarrow \theta_\omega = 45^\circ \tag{2.53}$$

which means the principle stiffness axes is 45° from the observing and tuning axes. At this angle, both modes experience the same stiffness contribution from the tuning electrode, and the relative frequency does not change. As the tuning increases, the principle stiffness axes are further deviated from the tuning axes. The opposite intrinsic mode becomes more aligned with the tuning electrode, and the two intrinsic modes become further apart (Fig. 2.7).

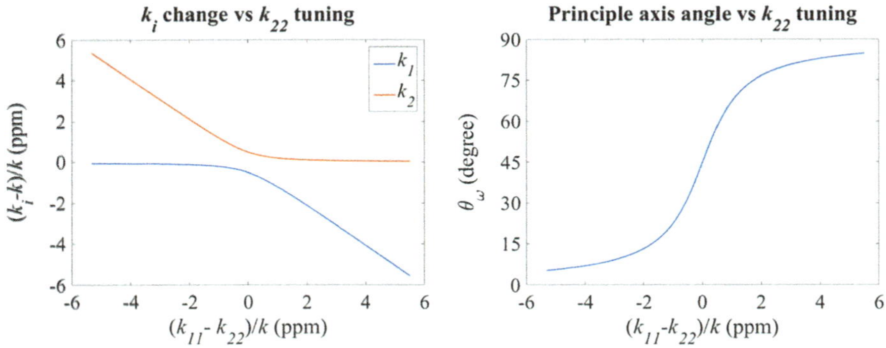

Fig. 2.7 Numerical calculation of principle stiffness and principle resonant axis misalignment with different k_{22} tuning conditions in the presence of a fixed k_{12}. Mode-matched operation can only be achieved when $k_1 = k_2$, which is prevented by veering effect

Therefore, cross-coupling (off-diagonal) terms in the observed stiffness matrix must be eliminated along with stiffness matching to enable true mode-matching.

2.2.5.2 Quadrature Cancellation

Quadrature tuning can be achieved in a similar manner as the stiffness mismatch tuning with the tuning electrode aligned in between the antinodes of the two orthogonal modes instead of aligned to the antinode of one of the modes.

For example, in a BAW disk, yaw gyroscope using orthogonal degenerate $n = 3$ wine-glass modes. The antinodes (maximum displacement amplitude) and nodes (minimum displacement amplitude) of the $n = 3$ wine-glass mode interchange around the circumference of the disk with separation angles of 30°. In the other mode, the antinodes and nodes are distributed in the same way but with locations exchanged. When looking at the mode shapes of both modes together, we can see the antinodes of the two modes are separated by 30°. An electrode placed at 15° from an antinode will have a capacitive gap that is equally affected by displacement of each mode:

$$\begin{aligned} C &= \frac{\varepsilon A}{[g_0 - \alpha(q_1 \pm q_2)]} = \frac{\varepsilon A}{g_0}\left\{1 + \frac{\alpha(q_1 \pm q_2)}{g_0} + \frac{\alpha^2(q_1^2 \pm 2q_1q_2 + q_2^2)}{g_0^2} + \ldots\right\} \\ \rightarrow \boldsymbol{K}_{\text{elec}} &= -\frac{\varepsilon A V_{\text{DC}}^2}{g_0^3}\alpha^2\begin{bmatrix} 1 & \pm 1 \\ \pm 1 & 1 \end{bmatrix}, \end{aligned} \tag{2.54}$$

where α is constant representing the ratio between the averaged displacement at the quadrature tuning electrode and the maximum displacement at the antinodes. The electrostatic stiffness matrix has identical diagonal terms therefore will not lead to relative frequency shift between the two modes. At the same time, the non-zero off-diagonal terms can be used to cancel the mechanical cross-coupling stiffness in

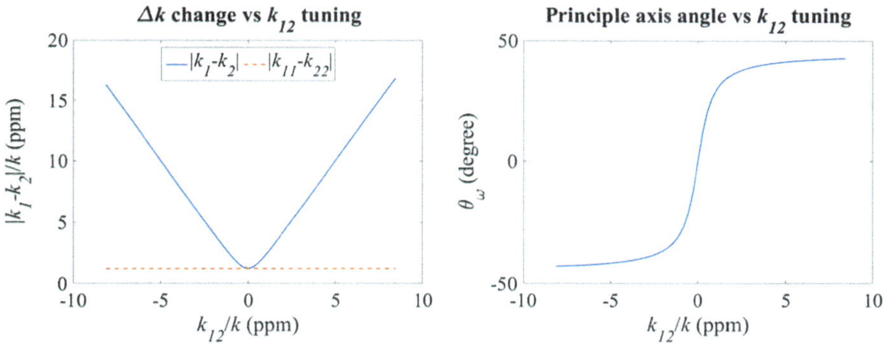

Fig. 2.8 Numerical calculation of principle stiffness mismatch and principle stiffness axis misalignment with different k_{12} tuning conditions in the presence of a fixed k_{11} and k_{22} mismatch. With optimum quadrature tuning, the principle stiffness axes become aligned with the observing axes, and the observed frequency split is determined by the mismatch between k_{11} and k_{22}

the gyroscope. In the principle-stiffness-axis coordinate view, the quadrature tuning reduces the mismatch between k_1 and k_2 and aligns the principle stiffness axes to the observing axes (Fig. 2.8).

In general, locations for quadrature tuning are easy to find or construct in MEMS yaw gyroscopes due to the two-dimensional nature of operational modes. However, as will be shown in the next chapter, quadrature tuning in MEMS pitch and roll gyroscopes is very challenging and critical, which requires novel solutions.

2.2.6 *Electrical Noise and Errors*

The electromechanical transduction and interface electronics have their own non-idealities, which contribute to the total noise and errors in the gyroscope system.

First, the interface electronics have intrinsic noise and errors including white noise, flicker noise, and drift errors. The high-frequency white noise directly contributes to the gyroscope output and limits the output resolution. The flicker noise is proportional to $1/f$ and is ideally low at the carrier frequency (i.e., the gyroscope resonant frequency). However, nonlinearity in both the MEMS and electronics will result in low-frequency noise up-conversion and cause flicker noise to show up in the demodulated output. Details related to nonlinear errors will be discussed in Chap. 4. Besides, discrete electronics are usually sensitive to environment variations such as temperature and stress variations, which will cause output in the output and limit the long-term stability of the gyroscope system.

Second, additional coupling between the drive channel and sense channel can occur in the electrical domain. A common coupling source in devices with capacitive transductions is the capacitive feedthrough. Feedthrough capacitance exists between nearby electrodes and signal lines with non-ideal shielding through which AC signal is coupled. The feedthrough impedance is given by $(j\omega_0 C_{FT})^{-1}$, where ω_0 is the angular frequency of the AC signal. Combining with (2.41) and (2.42), the overall output current bias is therefore:

$$i_{bias} = \left[-j \frac{c_{trans}\omega_0 d_{12}}{k_{22} - \omega_0^2 m_{22} + j\omega_0 d_{22}} + \frac{c_{trans}\left(\omega_0^2 m_{12} - k_{12}\right)}{k_{22} - \omega_0^2 m_{22} + j\omega_0 d_{22}} - j\omega_0 C_{FT} \right] v_d, \quad (2.55)$$

where c_{trans} represents the transduction constants in the sense current formulas. The feedthrough current is 90° out of phase with respect to the drive voltage. In mode-matched operation, the feedthrough current will be in quadrature phase with the rate signal, which may affect the accuracy of quadrature tuning and thus the gyroscope performance. Also, due to its capacitive nature, the feedthrough signal increases with frequency and affect the high-frequency gyroscopes more than their low-frequency counterpart. Therefore, proper routing design and feedthrough cancellation circuits [7] should be implemented in practice to reduce the effect of feedthrough.

Besides feedthrough, amplifiers in the interface circuit sharing the same ground and power lines can have cross talk. The effect of electronics cross talk is typically minimized by using a differential sensing configuration which rejects the cross-talk signal as a common mode signal. However, a large cross talk can still limit the stability and linearity of each individual amplifier. Therefore, proper circuit design should be used to minimize cross talk.

2.3 Performance Parameters

2.3.1 Scale-Factor and Bandwidth

A larger scale-factor is favorable for achieving higher signal-to-noise ratio (SNR) and suppressing the effect of electrical noise. From (2.44) we know that larger SF can be achieved by increasing the drive amplitude, sense-mode quality-factor, and the Coriolis coupling between the drive and sense modes.

Typically, the drive amplitude in parallel-plate resonators is limited to one-tenth the gap-size to ensure a linear transduction. While going to a smaller gap and comprising the drive amplitude are beneficial for the SF overall, there are ways to overcome this trade-off as will be shown in Chap. 4. The Coriolis coupling between the drive and sense modes (i.e., the Coriolis sensitivity) can be improved with design optimization. Especially for high-frequency pitch and roll gyroscopes with out-of-plane degrees of freedom, the conventional annulus design only provides a Coriolis-coupling coefficient of ~0.2; therefore there is still a large room to make improvement.

The bandwidth of a mode-matched gyroscope is evaluated by its sense-mode 3-dB bandwidth which is given by $f_0/(2Q_2)$, where f_0 is the sense-mode resonant frequency. Increasing the sense-mode quality-factor increases the SF, but it also leads to a smaller bandwidth. Therefore, a higher operation frequency is favorable to achieve a large bandwidth while having a high sense-mode quality-factor and scale-factor.

2.3.2 Angle Random Walk

The rotation rate output of a gyroscope needs to be integrated over time to obtain rotation angles and orientations. The random white noise in the system, when integrated over time, results in a zero-mean angle error with a standard deviation proportional to the square root of integration time. This deviation is described by the angle random walk (ARW) with a unit of $\left(^{\circ}/\sqrt{s}\right)$ or $\left(^{\circ}/\sqrt{h}\right)$. The ARW of a gyroscope can be evaluated by Allan variance measurement [8], and the value of ARW is given by the root Allan variance or Allan deviation (ADEV) value at first integration time, normally following a −1/2 slope line.

The white noise in the gyroscope system can also be expressed in the total noise equivalent rotation rate (TNEΩ) which has the unit of $\left(^{\circ}/s/\sqrt{\text{Hz}}\right)$ or $\left(^{\circ}/h/\sqrt{\text{Hz}}\right)$ and represents the resolution of the gyroscope per square root of detection bandwidth. The TNEΩ consists of two sources. First is the mechanical noise equivalent rate (MNEΩ) originated from the thermal noise-induced displacement in the mechanical structure of the gyroscope. At the sense-mode resonant frequency, the noise-induced sense displacement is given by:

$$q_{2_{\text{noise}}} = \sqrt{\frac{4k_B T Q_2}{m_{22}\omega_2^3}}, \tag{2.56}$$

where k_B is the Boltzmann constant and T is the ambient temperature. Comparing (2.56) and (2.34) gives the MNEΩ for mode-matched operation:

$$\text{MNE}\Omega = \frac{q_{2_{\text{noise}}}}{\text{SF}_{\text{mech}}} = \frac{1}{2\lambda|q_1|}\sqrt{\frac{4k_B T}{\omega_0 m_{22} Q_2}}. \tag{2.57}$$

Second, the white noise in the electronics will generate an electrical noise equivalent rate (ENEΩ), which is given by:

$$\text{ENE}\Omega = \frac{i_{s_{\text{noise}}}}{\text{SF}_{\text{elec}}}, \tag{2.58}$$

where $i_{s_{\text{noise}}}$ is the input-referred current noise density of the sense channel electronics. The two noise sources are uncorrelated, so the TNEΩ is given by:

$$\text{TNE}\Omega = \sqrt{\text{MNE}\Omega^2 + \text{MNE}\Omega^2}. \tag{2.59}$$

And the corresponding ARW is given by:

$$\text{ARW}\left({}^\circ/\sqrt{h}\right) = \frac{\text{TNE}\Omega\left({}^\circ/h/\sqrt{\text{Hz}}\right)}{60}. \tag{2.60}$$

From (2.57) and (2.58), we can see that both MNEΩ and ENEΩ can be reduced by increasing the SF. Therefore, implementing optimizations to increase SF as discussed in the previous section is important for achieving low ARW and high resolution in MEMS gyroscopes.

2.3.3 Bias Instability and Long-Term Drift

As given in (2.55), non-ideal MEMS gyroscope presents a bias current output without rotation input. The bias needs to be calibrated and subtracted before a gyroscope is used to obtain the actual rotation-induced output. However, any variations of the bias during the rotation measurement after the calibration can cause an error in the measured rotation rate. In an ideal mode-matched gyroscope, the quadrature-phase bias is eliminated by proper demodulation, leaving only the in-phase bias due to damping coupling.

During the calibration procedure, a demodulation phase is found to separate the in-phase channel and quadrature channel. During operation, this demodulation phase used is fixed; if there is a phase variation $\delta\varphi$ in the gyroscope mechanical response, there will be a corresponding demodulation phase error, and the demodulated output components become:

$$|i_{\text{demod}_i}| = |i_i \cos\delta\varphi + i_{\text{elec}_i}| + |i_q \sin\delta\varphi + i_{\text{elec}_q}| \tag{2.61}$$

$$|i_{\text{demod}_q}| = |i_q \cos\delta\varphi + i_{\text{elec}_q}| + |i_i \sin\delta\varphi + i_{\text{elec}_i}|, \tag{2.62}$$

where i_i and i_q are the actual instantaneous in-phase and quadrature components of the mechanical part given in (2.41) and (2.42) and $i_{\text{elec}i}$ and i_{elec_q} are the in-phase and quadrature components of the feedthrough and electrical coupling with respect to demodulation phase, respectively. For a gyroscope with identical drive and sense channel electronics, i_{elec_i} may come from electronics cross talk, and i_{elec_q} comes mainly from capacitive feedthrough. For a small phase error, we have:

$$|i_{\text{demod}_i}| = |i_i + i_{\text{elec}i}| + |i_q\delta\varphi| \tag{2.63}$$

$$|i_{\text{demod}_q}| = |i_q + i_{\text{elec}_q}| + |i_i\delta\varphi|. \tag{2.64}$$

From (2.63) we can see that the in-phase bias variation can potentially come from either variation in mechanical in-phase coupling level or quadrature injection due to demodulation errors. In practice, the demodulated quadrature is cancelled to close to zero through quadrature tuning, and demodulation error-caused bias variation will be small. However, as indicated in (2.64), with the presence of large feedthrough and other electrical coupling, making the demodulated quadrature signal zero does not correspond to minimum actual quadrature in the mechanical part, and the residual quadrature can lead to large demodulation error-caused bias variation.

In fact, many different factors can affect the coupling level as well as the demodulation phase, causing bias variations. First, examining the denominator of the gyroscope output bias terms in (2.55) reveals that the correct demodulation phase depends on both the frequency mismatch and the sense-mode dissipation. Therefore, variations in either one can lead to a demodulation error. Second, the in-phase coupling level is proportional to d_{12}, which depends on the damping symmetry between the drive and sense modes and the alignment of principle damping axes. Depending on the limiting damping mechanism, d_{12} may vary over time and cause bias drift. In addition, the bias level is proportional to the actuation voltage. As a result, any fluctuation in v_d will also translate into bias variations. In the electrical domain, noise and instability of the amplifiers and other instruments can also cause variations in the measured gyroscope output.

Bias errors can be sorted into two types. The first type is due to flicker noise in the gyroscope system. Since flicker noise is proportional to $1/f$, it does not reduce by averaging over time. Consequently, it shows up in the ADEV as a flat bias variation floor, which is called the bias instability (BI). Flicker noise can typically come from electrical noise up-conversion, noise in actuation voltage, and other instrument noise. There are also studies showing that the quantum limit of frequency stability of a resonator can present a $1/f$ behavior [9–11]. All these are potential causes of BI in MEMS gyroscope.

The other type of bias error is deterministic drift due to environmental effects which increases overtime and shows up in the ADEV as rate random walk (RRW) or rate ramp (RR). RRW and RR limit the long-term reliability of the gyroscopes and usually come from temperature- and stress-dependent parameters such as the sense-mode quality-factor and drift in electronics and measurement instruments.

Similar to error caused by thermal noise, the effective rate error caused by bias variation can be reduced with a larger SF. However, because the bias level and bias variations are usually proportional to v_d, increasing actuation level generally does not lead to significant improvements in bias equivalent rate stability. As a result, other methods to increase SF need to be used.

In summary, to achieve high performance in MEMS gyroscope, the scale-factor should be maximized to reduce the noise and bias equivalent rate output by improving the actuation range and Coriolis sensitivity, and the bias should be minimized and stabilized by cancelling quadrature and eliminating unstable damping mechanisms.

References

1. G.G. Coriolis, *Mémoire sur les équations du mouvement relatif des systèmes de corps* (Bachelier, 1835)
2. M.L. Foucault, Physical demonstration of the earth's motion of rotation, by means of the pendulum. London, Edinburgh, Dublin Philos. Mag. J. Sci **1**(7), 575–578 (1851)
3. M.F. Zaman et al., A mode-matched Silicon-Yaw Tuning-Fork gyroscope with subdegree-per-hour Allan deviation bias instability. J. Microelectromech. Syst. **17**(6), 1526–1536 (2008)
4. D. Lynch, Vibratory gyro analysis by the method of averaging, in *Proc. 2nd St. Petersburg Conf. on Gyroscopic Technology and Navigation*, (St. Petersburg, 1995), pp. 26–34
5. S. Vidoli, F. Vestroni, Veering phenomena in systems with gyroscopic coupling. J. Appl. Mech. **72**(5), 641–647 (2004)
6. B.J. Gallacher et al., Electrostatic correction of structural imperfections present in a microring gyroscope. J. Microelectron. Syst. **14**(2), 221–234 (2005)
7. C.L. Mayberry, *Interface Circuits for Readout and Control of a Micro-Hemispherical Resonating Gyroscope* (Georgia Institute of Technology, 2014)
8. N. El-Sheimy, H. Hou, X. Niu, Analysis and modeling of inertial sensors using Allan variance. IEEE Trans. Instrum. Meas. **57**(1), 140–149 (2008)
9. F.L. Walls et al., A new model of 1/f noise in BAW quartz resonators, in *Proceedings of the 1992 IEEE Frequency Control Symposium*, (1992), pp. 327–333
10. F. Sthal et al., Study on the origin of 1/ f noise in quartz resonators. J. Stat. Mech. Theory Exp **2016**(5), 054025 (2016)
11. M. Sansa et al., Frequency fluctuations in silicon nanoresonators. Nat. Nanotechnol. **11**, 552 (2016)

Chapter 3
Bias Control in Pitch and Roll Gyroscopes

Portable navigational applications require integrated high-performance triaxial rotation sensing. MEMS triaxial gyroscopes can be achieved by either assembling identical discrete single-axis gyroscopes in three orthogonal directions or using monolithic planar gyroscopes with different designs for each sensing axis on a single chip. The discrete approach subjects to orientation errors with temperature variation, mechanical shock, and other environment effects. In addition, the assembly involves size and cost penalties, which highly limit their feasible applications. Therefore, a single-chip approach is the more promising option for future IMUs. In order to achieve robust IMU for navigational applications, high performance is required for all three axes of rotation sensing. However, as mentioned in Chap. 1, achieving stable high-frequency resonant pitch or roll gyroscopes is challenging, and it has been the bottleneck of realizing robust MEMS IMUs for high-end applications. In this chapter, the major limits of bias level and bias stability in pitch and roll gyroscopes will be discussed, and effective solutions to overcome these limits will be presented.

3.1 Quadrature Cancellation

As pointed out in Chap. 2, quadrature cancellation is crucial for enabling mode-matching and reducing demodulation-related bias instability. In a single-chip approach, planar gyroscope designs with both IP and OOP DOF are used. In yaw gyroscopes with only IP modes, quadrature cancellation is realized by electrostatic quadrature tuning with electrodes orientated in between the antinode displacement directions of the two modes [1, 2]. However, in conventional pitch or roll gyroscopes using both IP and OOP modes, an effective location for implementing quadrature electrode is not available.

Pitch or roll gyroscopes fabricated by directional etching can have small trench tilting and tapering due to fabrication imperfections. An IP acoustic wave reflected

H. Wen, *Toward Inertial-Navigation-on-Chip*, Springer Theses,
https://doi.org/10.1007/978-3-030-25470-4_3

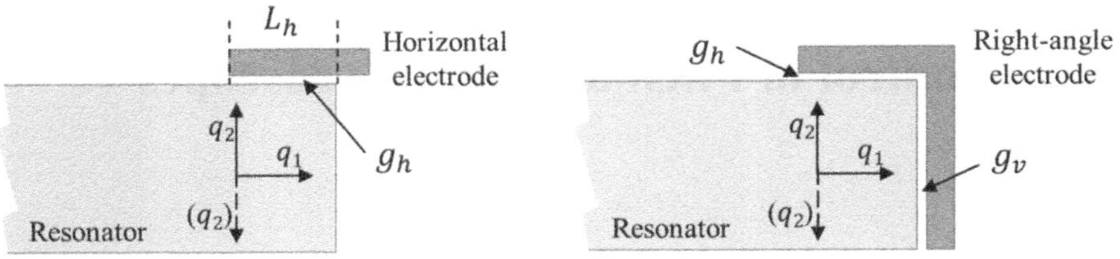

Fig. 3.1 Cross-sectional views of a horizontal electrode and a right-angle electrode in pitch or roll gyroscopes with both in-plane and out-of-plane motions [3]

from the tilted to boundary will have an OOP component, causing energy coupling between IP and OOP modes and leading to large quadrature errors. At the same time, the directional etching-based process can only provide electrodes with capacitive gaps in either horizontal or vertical directions, and achievable cross-coupling tuning with only horizontal, vertical, or combination of those two is very limited [3] (Fig. 3.1).

For example, assuming q_1 is in IP direction and q_2 is in OOP direction, the capacitance of a horizontally placed electrode will be:

$$C_h(q_1, q_2) = \frac{\varepsilon W_h L_h(q_1)}{g_h(q_2)} = \frac{\varepsilon W_h(L_{h0} + \alpha_1 q_1)}{g_{h0} \mp \alpha_2 q_2}, \tag{3.1}$$

where W_h, L_{h0}, and g_{h0} are the initial width, length of the overlapping area between the electrode and the resonant structure, and the initial gap-size when the gyroscope is not deforming. And α_1 and α_2 are mode-shape constants with values between 0 and 1 representing the ratio between the normal displacements at the electrode and the maximum displacements at the antinodes of each mode. According to (2.46), the electrostatic stiffness matrix provided by the horizontal electrode is:

$$\boldsymbol{K}_{\text{elec}} = \begin{bmatrix} k_{e11} & k_{e12} \\ k_{e12} & k_{e22} \end{bmatrix} \approx -\frac{\varepsilon W_h L_{h0} V_{\text{DC}}^2}{2 g_{h0}^3} \begin{bmatrix} 0 & \pm\alpha_1\alpha_2 \frac{g_{h0}}{L_{h0}} \\ \pm\alpha_1\alpha_2 \frac{g_{h0}}{L_{h0}} & 2\alpha_2^2 \end{bmatrix}. \tag{3.2}$$

Efficient tuning of stiff high-frequency gyroscopes relies on the use of large capacitance electrodes with large overlapping area and small gap-size. For MEMS gyroscopes operating in megahertz range, a L/g-ratio of a few hundred is typically needed for sufficient tuning. From (3.2) we can see the cross-axis tuning term k_{e12} provided by a horizontal electrode is orders of magnitude smaller than the typical OOP stiffness tuning term k_{e22} can offer. Therefore, the quadrature compensation, which can be achieved using horizontal electrode due to overlapping area change, is marginal and practically negligible compared to the typical quadrature levels in a non-ideal pitch or roll gyroscope. Similarly, a vertical electrode will be found to enable mainly IP stiffness tuning and negligible quadrature tuning. As a result, a combination of horizontal and vertical electrodes will also produce barely any quadrature tuning, even though intuitively combining IP and OOP springs seems to create an effective spring in between. For example, the capacitance of a right-angle electrode can be expressed as a separable function of q_1 and q_2 for tuning purpose:

$$\begin{aligned} C_r(q_1, q_2) &= C_v(q_1) + C_h(q_2) \\ &\approx \frac{\varepsilon}{g_0}\left[A_v + A_h + A_v\frac{\alpha_1 q_1}{g_0} + A_h\frac{\alpha_2 q_2}{g_0} + A_v\left(\frac{\alpha_1 q_1}{g_0}\right)^2 + A_h\left(\frac{\alpha_2 q_2}{g_0}\right)^2 + \ldots\right], \end{aligned} \tag{3.3}$$

where A_v and A_h are the overlapping areas between the resonant structure and the vertical and horizontal region of the electrode, respectively. The corresponding electrostatic stiffness matrix is:

$$\boldsymbol{K}_{\text{elec}} = \begin{bmatrix} k_{e11} & k_{e12} \\ k_{e12} & k_{e22} \end{bmatrix} \approx -\frac{\varepsilon V_{\text{DC}}^2}{g_0^3}\begin{bmatrix} \alpha_1^2 A_v & 0 \\ 0 & \alpha_2^2 A_h \end{bmatrix}. \tag{3.4}$$

The diagonal form of the stiffness matrix shows that a right-angle electrode can only effectively vary the stiffness in IP and OOP direction, but not the cross-coupling stiffness.

3.1.1 Slanted Electrode

To enable the cross-coupling tuning between IP and OOP modes, one needs to construct a capacitor with inseparable dependence on both IP and OOP displacement. An effective way to do this is by a parallel capacitor with a gap orientated at an angle from the horizontal or vertical direction. The MEMS implementation would be to create a gyroscope resonant structure with a slanted edge and put a nano-gap electrode along the slanted surface (Fig. 3.2). The capacitance of an electrode with slanting angle θ from the horizontal direction is given by:

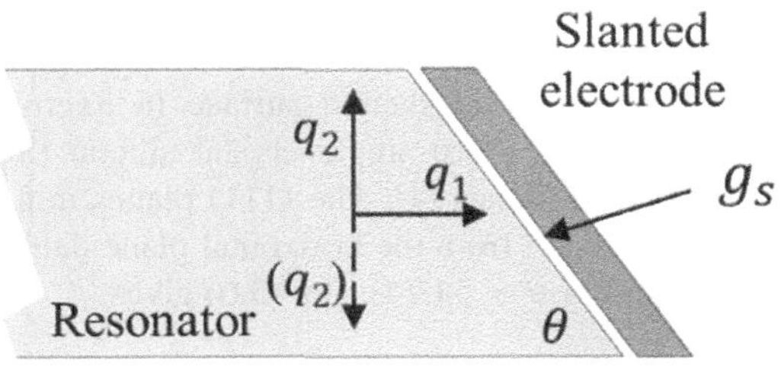

Fig. 3.2 Cross-sectional views of a slanted electrode in pitch or roll gyroscopes with both in-plane and out-of-plane motions [3]

$$C_s(q_1,q_2) = \frac{\varepsilon A_s}{g_s(q_1,q_2)} = \frac{\varepsilon A_s}{g_{s0} - (\alpha_1 q_1 \sin\theta \pm \alpha_2 q_2 \cos\theta)}$$
$$\approx \frac{\varepsilon A_s}{g_0}\left[1 + \frac{\alpha_1 q_1 \sin\theta \pm \alpha_2 q_2 \cos\theta}{g_0} + \left(\frac{\alpha_1 q_1 \sin\theta \pm \alpha_2 q_2 \cos\theta}{g_0}\right)^2 + \ldots\right]. \tag{3.5}$$

And the corresponding electrostatic stiffness matrix is calculated to be:

$$\boldsymbol{K}_{\text{elec}} = \begin{bmatrix} k_{e11} & k_{e12} \\ k_{e12} & k_{e22} \end{bmatrix}$$
$$\approx -\frac{\varepsilon A_s V_{\text{DC}}^2}{g_0^3}\begin{bmatrix} \alpha_1^2 \sin^2\theta & \pm\alpha_1\alpha_2 \sin\theta\cos\theta \\ \pm\alpha_1\alpha_2 \sin\theta\cos\theta & \alpha_2^2 \cos^2\theta \end{bmatrix}. \tag{3.6}$$

For an optimized slanted angle of 45°, Eq. (3.6) becomes:

$$\boldsymbol{K}_{\text{elec}} = \begin{bmatrix} k_{e11} & k_{e12} \\ k_{e12} & k_{e22} \end{bmatrix} \approx -\frac{\varepsilon A_s V_{\text{DC}}^2}{2g_0^3}\begin{bmatrix} \alpha_1^2 & \pm\alpha_1\alpha_2 \\ \pm\alpha_1\alpha_2 & \alpha_2^2 \end{bmatrix}. \tag{3.7}$$

Comparing (2.54) and (3.7) shows that high quadrature tuning efficiency similar to the quadrature tuning in yaw gyroscopes can be achieved in pitch or roll gyroscopes with slanted electrodes. When the slanting angle is not 45°, the slanted electrode will give a lower quadrature tuning efficiency. Since the diagonal terms in the stiffness matrix are not identical, it may also introduce a stiffness mismatch between the IP and OOP modes; therefore proper horizontal or vertical electrodes should be used in cooperation with the slanted electrodes during the mode-matching process.

3.1.2 *Fabrication Implementation*

The slanted electrode design enables efficient quadrature tuning in pitch and roll gyroscopes. However, creating the slanted electrode in micromachined gyroscopes is not a trivial task. In order to generate slanted profile with a proper slanting angle, etching technologies other than directional etching need to be involved. One practical way to create slanted surface in microfabrication is using the anisotropic wet-etching of (100) single crystal silicon (SCS), which is self-bounded by the (111) crystal planes [4]. The (111) planes in a (100) SCS wafers are slanted at an angle of 54.74° from the horizontal plane defined by the diamond cubic crystalline structure. For $\theta = 54.74^\circ$, Eq. (3.6) gives:

$$\boldsymbol{K}_{\text{elec}} = \begin{bmatrix} k_{e11} & k_{e12} \\ k_{e12} & k_{e22} \end{bmatrix} \approx -\frac{\varepsilon A_s V_{\text{DC}}^2}{g_0^3} \begin{bmatrix} 0.67\alpha_1^2 & \pm 0.47\alpha_1\alpha_2 \\ \pm 0.47\alpha_1\alpha_2 & 0.33\alpha_2^2 \end{bmatrix}. \tag{3.8}$$

which provides off-diagonal terms close to the optimum case given by (3.7).

Silicon anisotropic wet-etching is usually used for creating large cavities or releasing devices in microfabrication. Using it for precision fabrication of complicated MEMS gyroscope structures is challenging. As will be presented in the following sections, challenges like convex corner protection and integration with conventional fabrication process need to be addressed.

3.2 KOH-Etched Quad-Mass Gyroscope

3.2.1 Design of Wet-Etched-Only Gyroscope

The self-bounded anisotropic wet-etching provides etching time-insensitive structures with smooth slanted surfaces, giving it the potential for fabricating the resonant structure for MEMS gyroscopes. Besides enabling quadrature tuning in pitch or roll gyroscopes with OOP DOF, fabricating the resonant structure with anisotropic wet-etching can potentially offer immunity to the thickness variation of the wafer used to fabricate the gyroscope.

A resonant MEMS device requiring releasing and suspension is usually fabricated on a silicon-on-insulator (SOI) wafer, which consists of a SCS device layer where the resonant structure will be formed, a buried oxide (BOX) layer which will be partially removed to release the resonant structure, and a thick handle layer as a substrate for handling during processing and mounting of completed devices. Device layer thickness variation is one of the common non-idealities in SOI wafers. For typical commercial SOI wafers, the device layer thickness variation is ± 1 μm out of 40 μm. While wafers with smaller thickness variation may be available, the cost of such wafer is also much higher. The thickness variation can significantly affect the OOP frequencies of a MEMS resonator while introducing marginal changes to IP properties. In the case of a pitch of roll gyroscopes, this translates to a large frequency mismatch between the IP and OOP modes, which increases the tuning difficulties and limits the yield of mode-matched parts. The anisotropic wet-etching provides a possible solution to the thickness variation. The slanted structure relates the IP dimensions with the OOP thickness through the slanting angle. For example, a beam etched with anisotropic wet-etching on a (100) device layer will have an isosceles trapezoid cross section, with the top edge width defined by lithography and the nonparallel edges defined by the (111) planes. The base edge width (and therefore the averaged width) then becomes a function of the thickness:

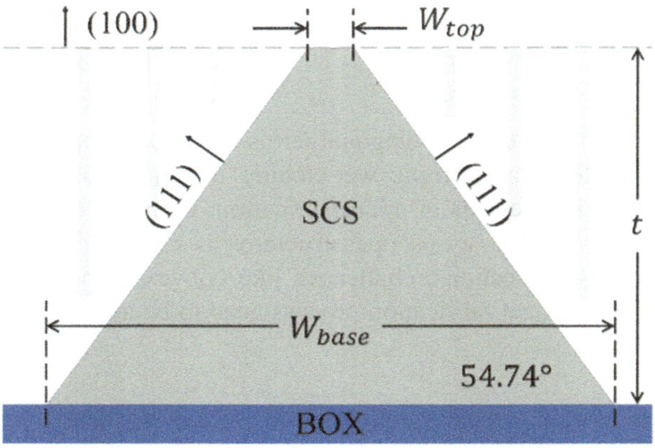

Fig. 3.3 Cross-sectional view of a trapezoidal beam created by anisotropic wet-etching of SCS. The width of the beam is coupled to its thickness through the slanting angle [3]

$$W_{\text{base}} = W_{\text{top}} + 2t\cot\theta \text{ and } W_{\text{average}} = W_{\text{top}} + t\cot\theta, \tag{3.9}$$

where t is the device layer thickness. The thickness dependence of the IP dimensions causes the IP and OOP resonant frequencies to follow a similar trend in the presence of thickness variations. In a gyroscope, this means the IP and OOP gyroscopic modes will track each other, and the frequency split will become thickness insensitive (Fig. 3.3).

In the past, silicon anisotropic wet-etching has been widely used for batch fabrication of passive devices like pressure sensors due to low processing cost. Therefore, replacing the more expensive dry etching process such as deep reactive-ion etching (DRIE) with anisotropic wet-etching in MEMS gyroscope fabrication will better suit for low-cost applications in the consumer market while enabling yield improvement and quadrature cancellation for better performance. Driven by these motivations, a pitch or roll gyroscope design that can be fabricated without DRIE and using only silicon anisotropic wet-etching is developed. A limitation of using only anisotropic wet-etching is that the resonant structure can only have straight outlines aligned to (110) or equivalent directions. To construct a pair of gyroscopic IP and OOP modes with balanced motions using allowed geometries, a quad-mass configuration is chosen, where four rectangular proofmass are connected through two pairs of straight beams. In the IP mode, the beams bend in IP directions and follow a Lamé-mode-like outline. Each proofmass vibrates with a rotating motion around the connection points following the bending of the beams. In the OOP mode, one pair of beams bends in OOP direction and the other pair of beams undergoes torsional deformations, causing the proofmass to rotate around axes along the torsional beams. These two modes are coupled through Coriolis force as indicated in Fig. 3.4, and dimensions of the proofmass and connecting beams can be adjusted to match the drive and sense mode frequency.

A prototype design using the presented gyroscopic modes is developed targeting SOI wafers with a 40-μm-thick device layer. The design possesses no vertical

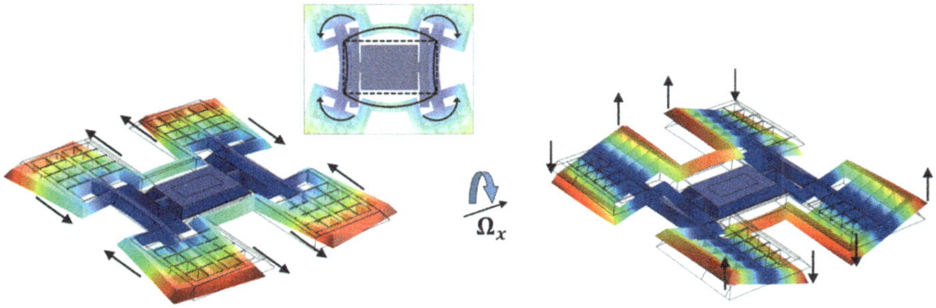

Fig. 3.4 Mode-shapes of a quad-mass roll gyroscope with structures feasible for wet-etching-only fabrication. The arrows indicate the motion of the proofmass in each mode and demonstrate the orthogonal relation and Coriolis coupling of the two modes

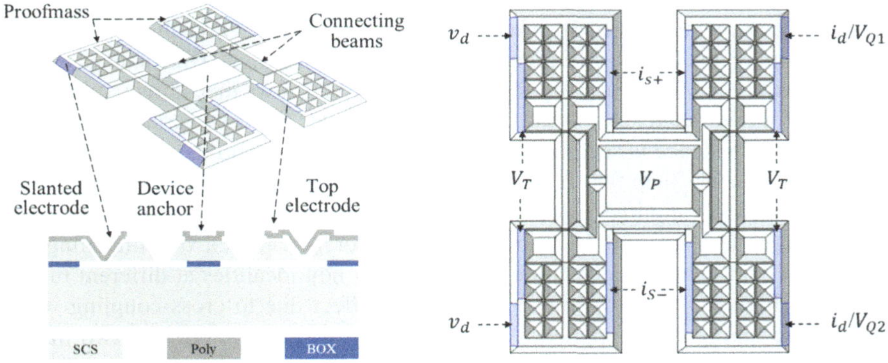

Fig. 3.5 Electrode cross-sectional view and electrical connection configuration. The purple region indicates location of the capacitive transduction. The reverse pyramid shapes in the proofmass are release holes formed by anisotropic wet-etching [3]

sidewall due to the absence of directional etching. Consequently, horizontal electrodes along the top surface of the resonant structure and slanted electrodes are used for the actuation, tuning, and readout of the gyroscope. As shown in Fig. 3.5, a pair of slanted electrodes is used to actuate the IP mode as drive mode. The actuation forces generated by the slanted electrodes have both IP and OOP components. However, the polarity of the displacement is symmetric at the two actuation electrodes for the IP mode and asymmetric for the OOP mode. Therefore, when the same actuation voltage is applied to the two actuation electrodes, the IP drive mode will be excited under the action of a common IP force, while the OOP sense mode excitation is avoided due to the lack of differential OOP force. The other pair of slanted electrodes is used for either drive mode readout or electrostatic quadrature tuning. As mentioned above, the IP and OOP modes have different polarity symmetries at the two slanted electrodes. From (3.8) we know, one of the slanted electrode will provide a positive k_{e12}; the other one will give a negative k_{e12}. Based on specific cases, one of the electrodes will be used for quadrature cancellation; the other one will be used for drive mode readout. Stiffness mismatch compensation (i.e., frequency tuning) and sense mode readout are done using top electrodes with horizontal gaps. Since only the sense mode frequency will be tuned using top electrode, the OOP mode should be designed to have higher untuned resonant frequencies, which can be brought down with electrostatic tuning to achieve mode-matched condition. And the tuning voltage used will be accommodating the process variations and material imperfections. The sensing output uses a differential configuration to reject common mode errors like electronics cross talk.

3.2.2 Simulation Results

Finite element method (FEM) simulations are done using COMSOL Multiphysics for developing and function verification of the prototype design. The finalized design has an operational frequency of ~200 kHz. Figure 3.6 shows the simulated IP mode and OOP mode frequencies with ±1 μm device layer thickness variations. A clear thickness dependence is seen for the IP mode as expected, and the total frequency-split change is less than 900 Hz across the thickness variation. Tuning coverage of common process variations is found to be achievable with ~500 nm tuning gap using DC voltages less than 20 V. The tuning behavior is captured by simulating the gyroscope with intentionally introduced geometry non-idealities at different tuning voltages. Figure 3.7 clearly shows the veering effect due to cross-coupling when quadrature tuning is not applied. At the minimum frequency split, the two modes become mixed and indistinguishable for tuning purpose, which corresponds to a principle axis misalignment of 45° as described in (2.53). When proper quadrature tuning voltage is applied to the slanted electrode, the off-diagonal electrostatic stiffness can effectively cancel the mechanical cross-coupling stiffness terms, and decoupling between the IP and OOP modes can be achieved. When quadrature is

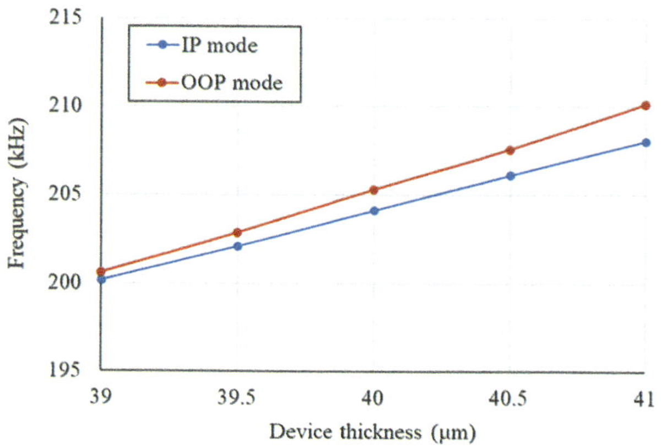

Fig. 3.6 Frequencies of IP and OOP modes across thickness variation [3]

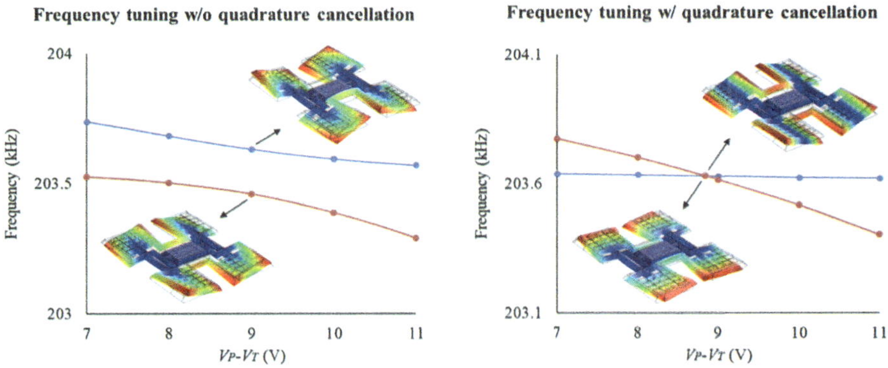

Fig. 3.7 Tuning curves of the quad-mass pitch or roll gyroscope with and without quadrature cancellation using slanted electrodes [3]

cancelled, the principle axes become aligned to the tuning axes; therefore the OOP mode can be tuned independently and mode-matching can be achieved.

3.2.3 Fabrication Process

The biggest challenge in fabricating resonant structures with anisotropic wet-etching is the convex corner undercut. At convex corners where two (111) planes meet, higher-order crystal planes which do not stop the etching become accessible to the etchant. As a result, undercut will occur at the convex corners of the etching mask. One way to mitigate the effect of the undercut is by extending the etching mask at the convex corners to cover sacrificial regions which will be consumed during the wet-etching and leave a close-to-desired final geometry. However, this method is etching time sensitive and not suitable for consistent batch fabrication. In addition, the undercut leaves large regions of suspended mask layers, affecting process steps following the wet-etching. A more suitable solution is using a two-mask local oxidation of silicon (LOCOS) method [5]. This method is time insensitive and can perfectly preserve the convex corner structures. However, it subjects to misalignment errors between the two wet-etching masks used. To avoid the misalignment error, a self-aligned two-mask LOCOS method is developed. In the self-aligned method, the entire trench pattern defining the resonator geometry is included on a first thick nitride mask layer. A second thin nitride mask layer is added on top of the first mask to partially reveal the trench pattern without exposing the convex corner regions. The patterns on the second mask are widened at non-convex-corner regions. When the wet-etching is performed, the edge of the trench will be defined by the first mask instead of the second mask as the etching propagates from the widened region along the (111) planes. Therefore, the finished trench profile is independent of the second mask and insensitive to misalignment between the two

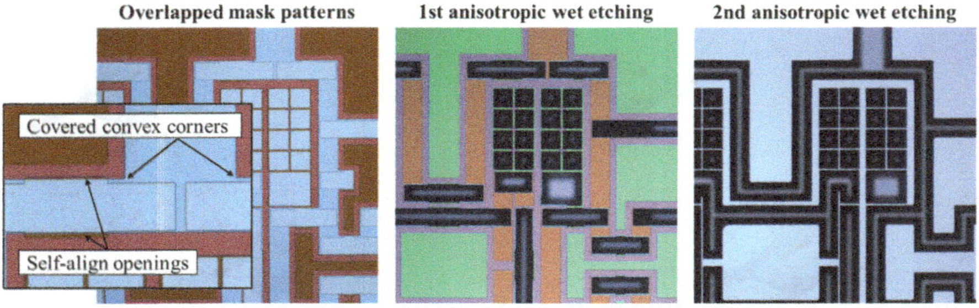

Fig. 3.8 Images of wafer processed with the self-aligned two-mask wet-etching process. Color differences of the mask layers between left and middle figures are due to different mask thickness on the different samples used to obtain the pictures

masks. After the first wet-etching, local oxidation is performed to protect the (111) surfaces exposed. The thin nitride mask is then removed, and a second wet-etching is performed to form the final structure with intact convex corners (Fig. 3.8).

Slanted and top electrodes with submicron capacitive gaps are formed using a polysilicon and sacrificial oxide method, where a thin sacrificial oxide (SACOX) is used to define the gap-size and polysilicon is used to as the electrode material. Details of the process flow can be found in Appendix A.1.

3.2.4 Experimental Characterization

The prototype gyroscope is fabricated on a 40 μm (100) SOI wafer. 45% potassium hydroxide (KOH) solution is used for the wet-etching for a high (100) to (111) selectivity. Figure 3.9 shows the SEM pictures of a fabricated gyroscope. Polysilicon slanted electrodes and top electrodes are formed with 540 nm and 510 nm capacitive gaps, respectively.

3.2.4.1 Mode-Matching Behavior

The fabricated gyroscope is wire-bonded to a testing PCB board and characterized inside a vacuum chamber. To verify the functionality of the slanted quadrature tuning electrodes and the mode-matching capability, the frequency response of the gyroscope is captured using a 2-channel 4-port network analyzer. The IP mode is actuated and recorded using channel 1 (S21), and the OOP mode is actuated and recorded using channel 2 (S43).

The OOP mode is found to be ~250 Hz lower than the IP mode when a polarization voltage of $V_P = 19$ V is used, and all tuning electrodes are kept at 0 V. When the tuning voltage VT on the top electrode is increased, the DC voltage difference between the resonator and the electrode reduces. Therefore, the OOP

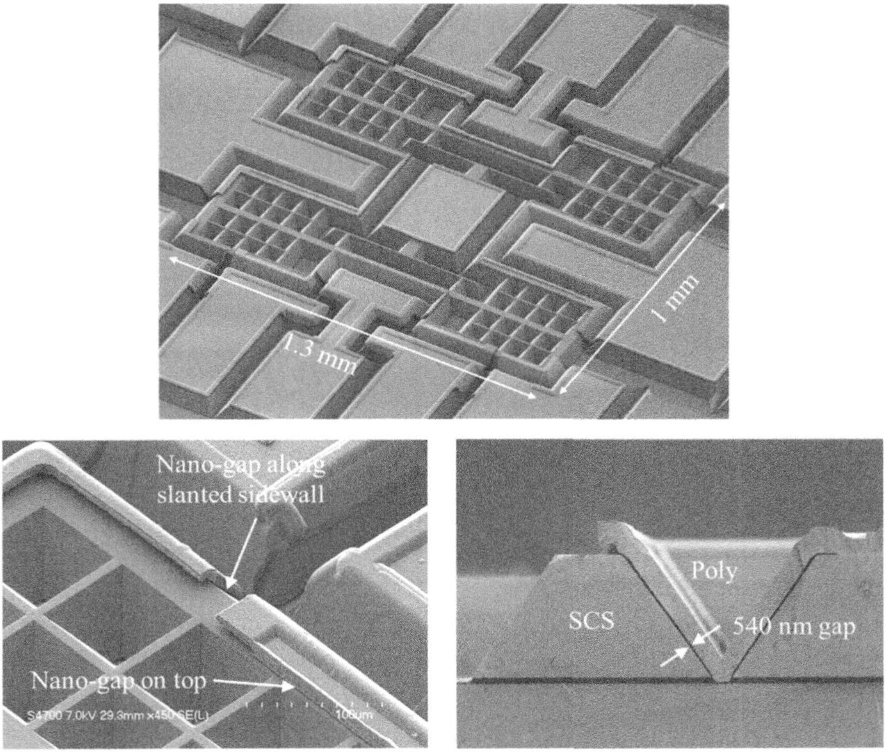

Fig. 3.9 SEM images of fabricated quad-mass pitch or roll gyroscope [3]. Cross-sectional view shows the slanted electrode with 540 nm capacitive gap

mode frequency starts to rise and the frequency split becomes smaller. However, a clear veering effect is found, causing a minimum frequency split of 64 Hz (Fig. 3.10).

The quadrature level is effectively reduced by increasing the quadrature tuning voltage on the proper slanted electrode. The cross-coupling peaks due to misalignment between the principle axes, and the observing axes in both channels are efficiently reduced, with the minimum level achieved with $V_{Q1} = 6.86$ V. With the optimized quadrature tuning voltage, the veering phenomenon is eliminated, and the OOP mode can be tuned independently. Mode-matching is then achieved with $V_T = 1.416$ V. The mode-matched peaks show Q of 21k for the IP mode and 27k for the OOP mode (Fig. 3.11).

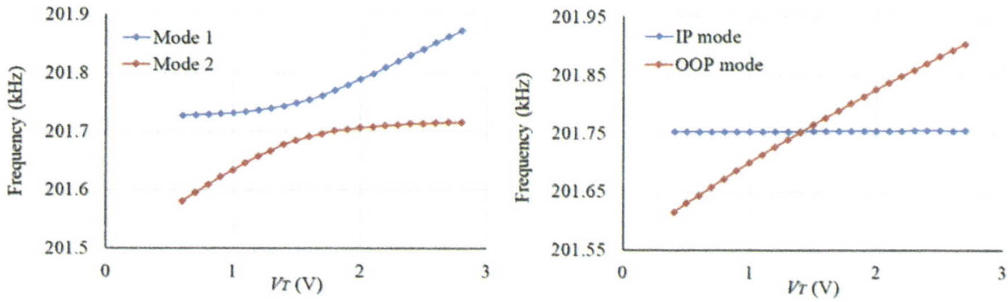

Fig. 3.10 Measured frequency tuning curve without (left) and with (right) optimized quadrature tuning [3]

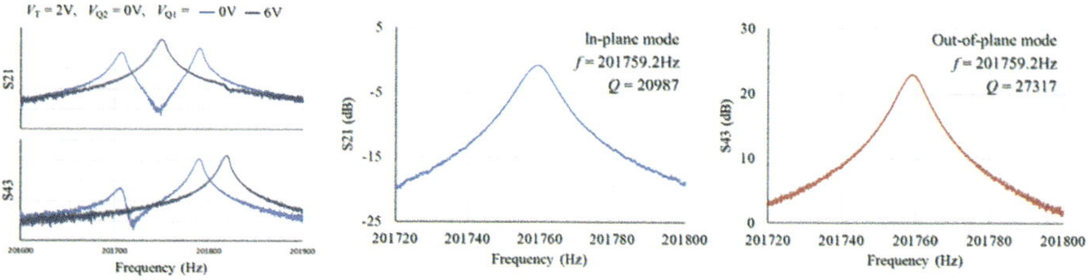

Fig. 3.11 (Left) The effect of quadrature cancellation when V_T is fixed at 2 V, and the quadrature tuning voltage V_{Q1} is adjusted. And mode-matched IP drive mode (middle) and OOP sense mode (right) [3]

3.2.4.2 Gyroscope Performance Characterization

The device is interfaced with an HF2LI lock-in amplifier and characterized as a gyroscope on a rotation rate table. Feedthrough cancellation circuit is added to reduce the electrical coupling between input and output. To evaluate the effect of quadrature cancellation, the quadrature tuning voltages are first kept at 0 V, and the two modes are tuning to the minimum frequency split. The rotation rate response is tested which gives a scale factor of 3.42pA/(°/s). The gyroscope is then mode-matched with the quadrature tuning voltage at the optimum value. The rotation rate response is recorded again, which shows a SF of 64.4pA/(°/s), offering ~20× improvement in SF and SNR (Fig. 3.12).

The noise performance of the resonant gyroscope is measured, and the ADEV is plotted in Fig. 3.13. An ARW is read at 1 s averaging time to be 1 °/√h, with a fitted value of 0.63°/√h following the −1/2 slope. The irregular bump around 0.017 s is suspected to be caused by the 60 Hz power line interference. The BI of the resonant gyroscope is found to be of 37.7°/h, which is more than 20× better than the best result reported for high-frequency pitch or roll gyroscopes in previous studies. Table 3.1 summarizes the specifications and measured performance of the gyroscope.

By implementing the slanted electrodes, the anisotropically wet-etched gyroscope successfully demonstrated the electrostatic quadrature cancellation in MEMS pitch

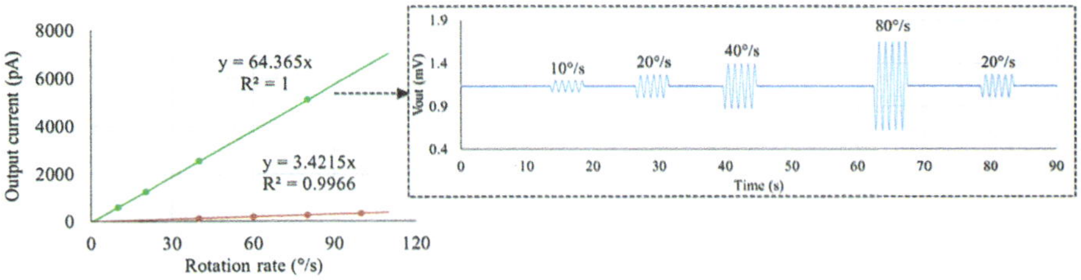

Fig. 3.12 Rate response of the mode-matched resonant gyroscope and the comparison between measured SF for quadrature-cancelled mode-matched operation (green) and minimum frequency-split operation without quadrature tuning (red) [3]

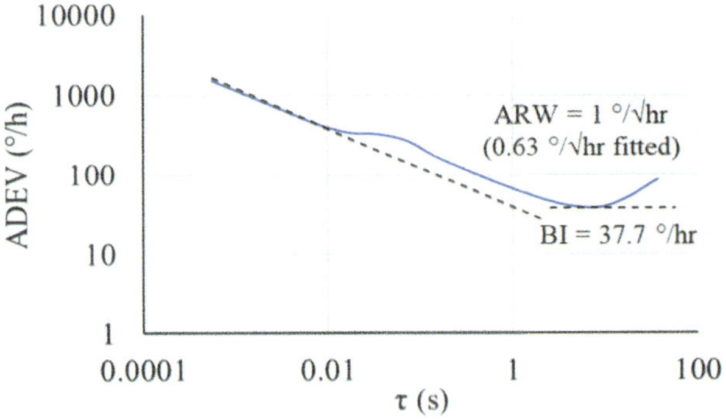

Fig. 3.13 Measured Allan deviation of the mode-matched quad-mass pitch or gyroscope [3]

Table 3.1 Performance metrics and specifications summary of the quad-mass resonant pitch or roll gyroscope [3]

Parameter	Value	Unit
Resonator size	1 × 1.3	mm × mm
Top gap-size	510	nm
Slanted gap-size	540	nm
Resonant frequency	201.7592	kHz
Mode-matched Q	21,000 (drive) 27,000 (sense)	
Minimum frequency split	64 (w/o V_Q) <0.1 (w/ V_Q)	Hz
Scale factor	3.42 (w/o V_Q) 64.4 (w/ V_Q)	pA/(°/s)
ARW	0.63	°/√h
Bias instability	37.7	°/h

or roll gyroscopes for the first time. And with the quadrature cancellation, the bias stability of the gyroscope is found to be improved significantly. As mentioned at the beginning of this chapter, realizing high performance through the low-cost wet-etching process is of great value, and the resonant quad-mass pitch or roll gyroscope can be found useful in many consumer applications.

However, the low-cost fabrication process also leads to some limitations. As mentioned previously, KOH solution is chosen as the etchant for the gyroscope fabrication because it has a high (100)-to-(111) etch selectivity, which allows for a better control of the final geometry can be achieved. But the drawback of KOH is its incompatibility with standard CMOS process. Therefore, integration of a KOH-processed gyroscope with interface ASIC becomes challenging and may lead to additional size and cost penalties. In addition, KOH also etches oxide at a slow rate. As a result, the oxide layer used in the LOCOS process may not survive very long etching. For thick devices requiring long etching, this will lead to non-ideal etching result at the convex corners (Fig. 3.14).

Design-wise, the anisotropic wet-etching cannot produce thin flexures for substrate-decoupling. To avoid a large anchor loss which will lead to unstable bias from damping coupling, a center support without rigid connections to the proofmass is used in the quad-mass pitch or gyroscope. This results in a limited OOP stability with low-frequency modes in the range of 20–50 kHz, which can cause poor robustness against shock and random vibrations (Fig. 3.15). Therefore, implementing the slanted electrodes and quadrature tuning in a robust high-frequency gyroscope becomes more valuable for applications with higher reliability requirements such as navigational applications.

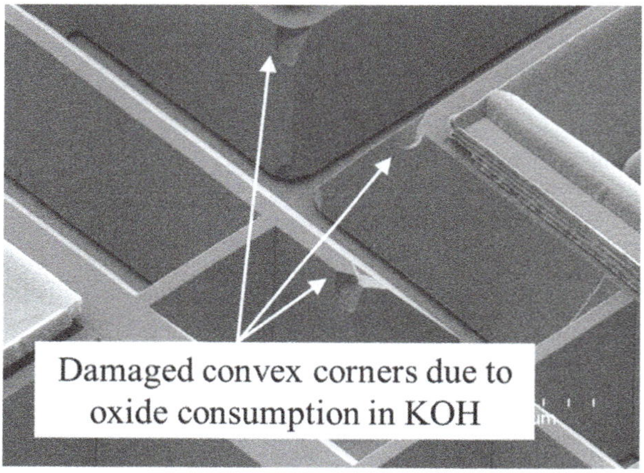

Fig. 3.14 Convex corner damage due to oxide protection layer consumed by KOH during long anisotropic wet-etching [3]

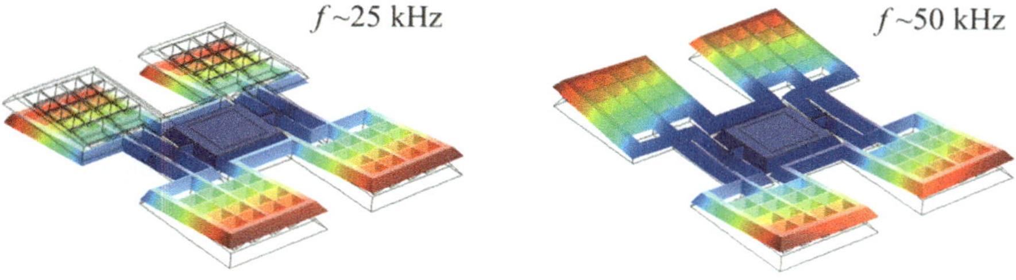

Fig. 3.15 Low-frequency OOP resonant modes of the quad-mass gyroscope

3.3 Dissipation and In-Phase Bias Stabilization

As discussed in Chap. 2, the in-phase bias component is indistinguishable from Coriolis force-induced signal and cannot be rejected through phase demodulation. Therefore, erratic changes in the in-phase bias level must be minimized to enable a high bias stability. The in-phase bias equivalent rotation rate can be derived from (2.35) as:

$$B_{\Omega_{\text{in-phase}}} = \frac{-d_{12}}{2\lambda m_{22}} \propto \Delta d \sin 2\theta_{\tau}. \tag{3.10}$$

Equation (3.10) shows the in-phase bias level of a gyroscope that is closely related to the damping mismatch between the two operation modes and the principle damping axis misalignment. Therefore, designing a gyroscope to have drive and sense modes with either constant or predictable dissipations is very important.

A measurable parameter that can be used to evaluate the damping factor of a resonant mode is its Q-factor. The Q-factor of a MEMS resonator depends on many different dissipation mechanisms. The Q limit placed by each dissipation mechanism is inversely proportional to the corresponding damping factor. The total dissipation is given by the sum of all different damping factors, and overall Q is given by:

$$\frac{1}{Q_{\text{total}}} = \frac{1}{Q_{\text{surface}}} + \frac{1}{Q_{\text{TED}}} + \frac{1}{Q_{\text{SFD}}} + \frac{1}{Q_{\text{anchor}}} + \frac{1}{Q_{\text{Akheiser}}} + \ldots \tag{3.11}$$

The equation above lists the Q limits due to dissipation mechanisms typically seen in MEMS resonators. Q_{surface} is related to the scattering loss at rough surfaces or interface between different materials. When using high-frequency bulk mode with capacitive transduction, the acoustic energy is mostly stored in the bulk structure of the resonator, and surface loss contribution will be minimal.

Q_{TED} corresponds to thermoelastic damping (TED) caused by mechanical thermal coupling through the coefficient of thermal expansion. During vibration, mechanical contraction and expansion introduce a temperature gradient in the resonator which leads to energy dissipation through irreversible heat transfer. Analytical expression is available for TED, which gives a well-defined temperature dependence and is independent of other environment variables [6]. Therefore, Q variations due to TED can usually be compensated by calibration or eliminated by temperature control using chip-level ovenization.

Q_{SFD} corresponds to squeeze-film damping (SFD), and in some cases slide film damping, caused by the energy exchange between the resonator and the gas trapped between the capacitive gaps. SFD is a function of the gas cutoff frequency which depends on the ambient pressure. For high-frequency devices with vacuum package, the resonance frequency is usually higher than the cutoff frequency, and the SFD is much smaller compared to that for low-frequency devices. However, SFD is also proportional to g^{-3}, where g is the capacitive gap-size. As we move to smaller gap-size, SFD may start to contribute to the overall Q limit. Also, SFD could be a source of damping axis misalignment, which is the capacitive gap-size is not uniform due to hydrodynamic lift effect. Nevertheless, SFD can also be expressed analytically and has well-defined temperature behavior. Therefore, even if SFD is the dominant dissipation mechanism, its effect can be calibrated and compensated in a similar fashion as TED.

Q_{anchor} corresponds to anchor loss, which is due to elastic energy leakage through anchoring structure to the substrate. The elastic wave propagation highly depends on the stress distribution at the anchor. This distribution can be affected by many different variables such as thermal stress of the substrate and linear acceleration induced stress. The change of stress distribution differs from mode to mode and is not necessarily monotonic. Therefore, anchor loss can cause erratic changes in both damping mismatch and principle damping axis misalignment. To achieve high bias stability, anchor loss of the gyroscopic modes must be minimized. At the same time, the anchoring structure should be robust to provide enough resistance against shock and vibration for consistent performance delivery.

Q_{Akheiser} represents the Akheiser limit of the material, which is the quantum limit of the fQ-product for a resonator made from a certain material. For SCS silicon, the fQ-product limit is ~2×10^{13} [7, 8]. With typical operation frequency below 10 MHz, the Akheiser limit in silicon MEMS gyroscope is usually not the dominating factor.

Besides controlling the Q behavior for in-phase bias stabilization, increasing the overall Q will reduce the mechanical noise and improve the SF, therefore enabling ARW meeting navigational application requirements.

3.4 Robust Substrate-Decoupled Pitch or Roll Gyroscope

High-frequency resonant gyroscopes fabricated using DRIE-based HARPSS™ process are known to have the capability of providing high vibration resistance and shock survivability as well as low anchor loss with proper substrate-decoupled designs. HARPSS™ process can be used to fabricate both yaw and pitch/roll high-frequency gyroscopes, but the yield and performance of high-frequency resonant pitch or roll gyroscopes are highly limited by quadrature errors. Therefore, implementing slanted electrode quadrature tuning technology in high-frequency pitch or roll gyroscopes is highly desirable.

3.4.1 Slanted Electrode in DRIE-Based Devices

Slanted surfaces are generated through anisotropic wet-etching of SCS. However, the anisotropic wet-etching is not compatible with traditional DRIE process and fabrication of integrated IMU. To solve this problem, a HARPSS plus wet-etching

(HARPSS+) process is developed, which combines silicon anisotropic wet-etching with DRIE process and allows the integration of multi-axis high-frequency resonant gyroscopes on a single chip [9].

The integration of anisotropic wet-etching with DRIE is done through vertical trench protection with refilled LPCVD tetraethyl orthosilicate (TEOS). First, DRIE trenches are etched through the device layer of a (100) SOI wafer, followed by steps needed to form vertical polysilicon electrodes as described in HARPSS™ process. The DRIE trenches are then refilled with LPCVD TEOS to protect the vertical sidewalls to be prepared for anisotropic wet-etching. A layer of TEOS is deposited and patterned on the top surface of the wafer as mask layer for anisotropic wet-etching. The DRIE trench pattern is especially designed at the locations where slanted surfaces are to be formed. Figure 3.16 shows the schematics of the wet-etching steps. The wet-etching mask has a "T"-shape pattern with one side aligned to the DRIE trenches. On the resonator side, the opening pattern forms a concave region, and on the electrode side, the pattern overlaps with the TEOS-filled trenches and forms a convex outline.

During the wet-etching, (100) plane is etched down and (111) surfaces will be exposed. On the device side, three (111) planes form a concave shape and stop the lateral etching. On the electrode side, (111) planes meet at two convex corners, which will not stop the wet-etching. Eventually, silicon will be fully etched along the edges of the convex corners, leaving a final structure defined by the concave (111) silicon plane, refilled vertical sidewall, and the BOX layer. After the wet-etching, oxidation and polysilicon surface micromachining steps can be used to form slanted and top electrodes. The complete HARPSS+ process flow is described in Appendix A.2.

HARPSS+-fabricated high-frequency devices have the main mechanical structure defined by DRIE instead of anisotropic wet-etching; therefore more complicated geometries such as annulus with curved outlines and thin beam networks can be implemented, making it suitable for multi-axis sensor integration. In addition, the

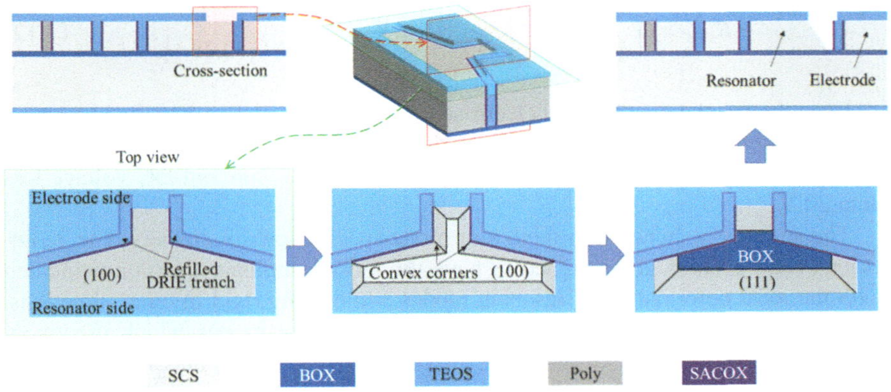

Fig. 3.16 Slanted surface generation steps in the HARPSS+ process flow [9]

wet-etching only serves to provide slanted surface for tuning purpose and does not affect the mechanical property of the resonator significantly. Consequently, larger (111) plane etching can be tolerated, and (100) to (111) selectivity becomes less critical. Therefore, a COMS compatible tetramethylammonium hydroxide (TMAH) etchant can be used instead of KOH, making the MEMS and ASIC integration less challenging.

3.4.2 High-Frequency Annulus Gyroscope with Slanted Electrodes

To verify the feasibility of HARPSS+ process and quadrature tuning in high-frequency pitch or roll gyroscope, an annulus gyroscope is designed with slanted electrodes incorporated. The design uses the conventional $n = 2$ IP wine-glass mode and $n = 3$ OOP bending mode as drive mode and sense mode, respectively, with a high operation frequency of 0.53 MHz. Slanted electrodes are built at the outer edge of the annulus. The ideal location for placing slanted electrode is at the coincided antinodes of the drive and sense modes, which gives mode-shape constants $\alpha_1 = \alpha_2 = 1$ as defined in Sect. 3.1 and maximized tuning efficiency. However, the antinodes of sense mode are also optimum locations for signal readout and OOP frequency tuning. Occupying the antinodes with slanted electrodes will sacrifice the frequency tuning range needed for frequency mismatch compensation. As mentioned before, for pitch or roll gyroscopes using both IP and OOP non-degenerate modes, the device thickness variation can cause a large frequency split, and the frequency mismatch usually requires more electrostatic compensation than quadrature. Therefore, the slanted electrodes are designed to be at a compromised location near the nodes of the IP mode, with the displacement constants simulated to be $\alpha_1 = 0.24$ and $\alpha_2 = 0.71$. Based on (3.8):

$$\boldsymbol{K}_{\text{elec}} = \begin{bmatrix} k_{e11} & k_{e12} \\ k_{e12} & k_{e22} \end{bmatrix} \approx -\frac{\varepsilon A_s V_{\text{DC}}^2}{g_0^3} \begin{bmatrix} 0.04 & \pm 0.08 \\ \pm 0.08 & 0.16 \end{bmatrix}. \tag{3.12}$$

Even though the off-diagonal electrostatic stiffness is smaller than the ideal case, it is still found in simulation to be sufficient to cover quadrature due to a typical $\pm 0.3°$ DRIE trench tilting errors with a gap -size of 300 nm and DC voltage less than 20 V.

The annulus is anchored through four mode-specific low-loss tethers. The geometries of the anchors are designed to have the most compliant direction coincided with the local deformation of resonant modes while being rigid in other directions. As a result, the tethers appear to be floppy for the operation modes, which effectively isolates the high-frequency acoustic waves from the anchor and decouples the resonant structure from the substrate, ensuring a low anchor loss and more stable damping coupling (Fig. 3.17).

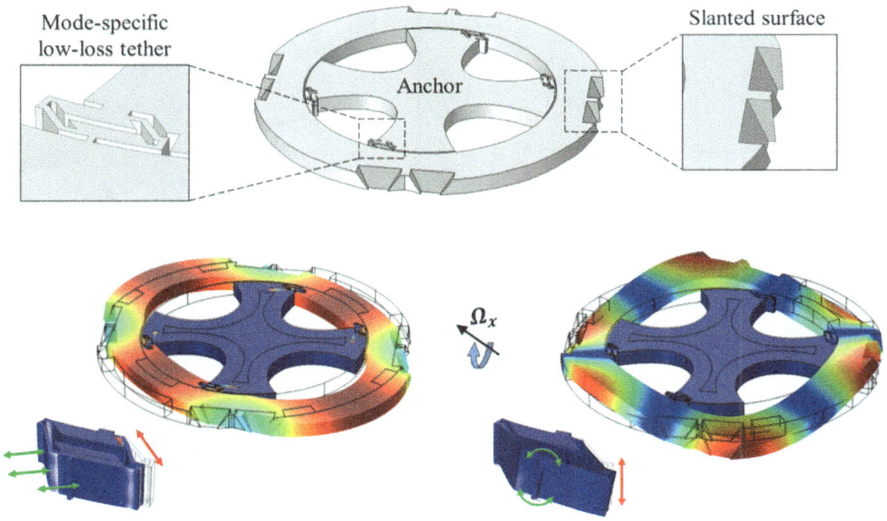

Fig. 3.17 Annulus gyroscope with slanted surface and mode-specific low-loss tether design. The anchors allow the motion of the operation modes (green arrows) but reject translation motions (red arrows) due to linear vibration [9]

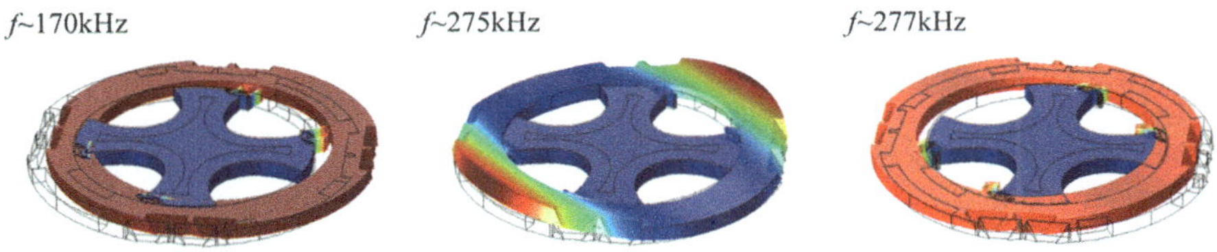

Fig. 3.18 Translational modes of the annulus with mode-specific tether design

At the same time, the tethers stay rigid for other degrees-of-freedom, and the combination of four tethers with different complaint directions provides a good translational motion rejection. All translational modes of the structure have frequencies above 170 kHz, which promises a high robustness against shock and vibration (Fig. 3.18).

3.4.3 Experimental Characterization

The high-frequency annulus pitch or roll gyroscope with slanted electrodes is implemented on a 60 μm SOI wafer using the HARPSS+ process. Electrodes with ~300 nm capacitive gaps are successfully formed in all three orientations including the vertical electrode, top electrode, and slanted electrode (Fig. 3.19).

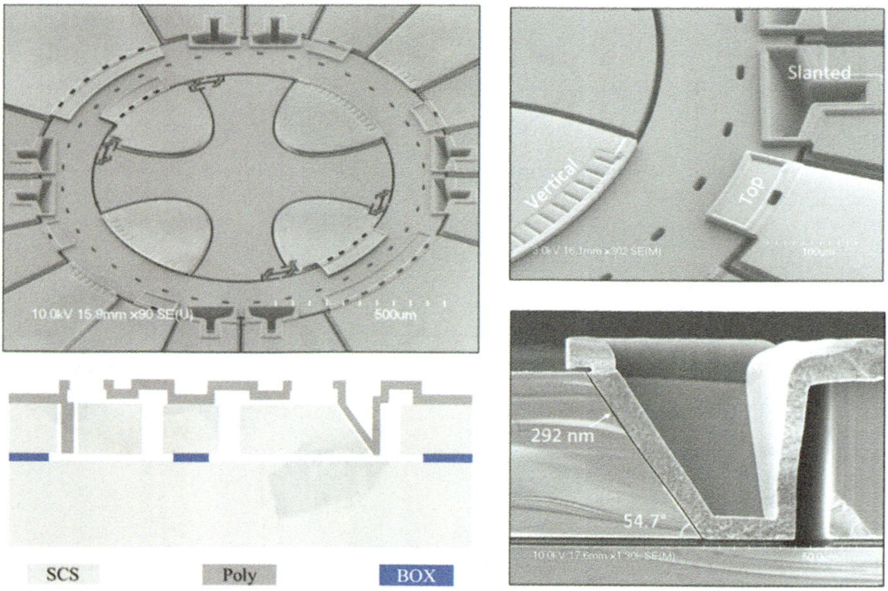

Fig. 3.19 SEM images of fabricated annulus gyroscope with slanted electrodes and schematics showing the cross-sectional view of different types of electrodes [9]

3.4.3.1 Resonant Behavior

The quadrature cancellation and mode-matching of the annulus gyroscope are characterized in a vacuum chamber with a two-channel four-port network analyzer. Channel 1 is used for the IP mode actuation and readout (S21), and channel 2 is used for the OOP mode actuation and readout (S43). The indirect response of the sense mode (S41) is also recorded where actuation is only applied to the IP mode and signal is read out from the OOP mode. Without quadrature tuning, two peaks are observed in each channel due to the cross-coupling between IP and OOP modes. And clear quadrature level can be seen in the indirect response. When quadrature tuning voltage is applied to the proper slanted electrodes, the minimum frequency split between the two modes becomes smaller, and the quadrature level seen in the indirect response is reduced, which verified the functionality of slanted electrode in high-frequency pitch or roll gyroscopes. By fine-tuning the frequency and quadrature tuning voltages, quadrature-cancelled mode-matching is achieved with optimized operation voltages being $V_P = 19$ V, $V_{T2} = 1.28$ V, $V_{Q2} = 1.55$ V, and $V_{T1} = V_{Q1} = 0$ V (Fig. 3.20).

As mentioned before, the HARPSS+ process enables simultaneous fabrication of high-frequency multi-axis resonant gyroscopes. This is verified by measuring an IMU die fabricated using the HARPSS+ process including both a BAW disk yaw gyroscope and pitch or roll annulus gyroscopes with slanted electrodes, which shows

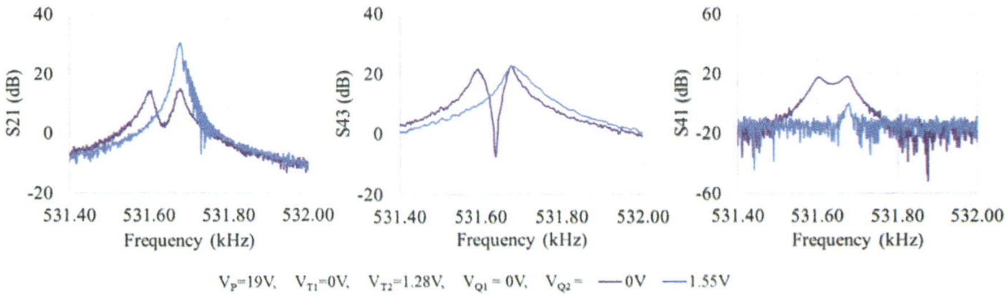

Fig. 3.20 Quadrature tuning and mode-matching of the annulus gyroscope. The IP and OOP modes are matched at ~0.53 MHz with Q of 118k and 18k, respectively. The ripples in the mode-matched drive peak come from the ring-down behavior due to the high Q and a fast sweeping time [9]

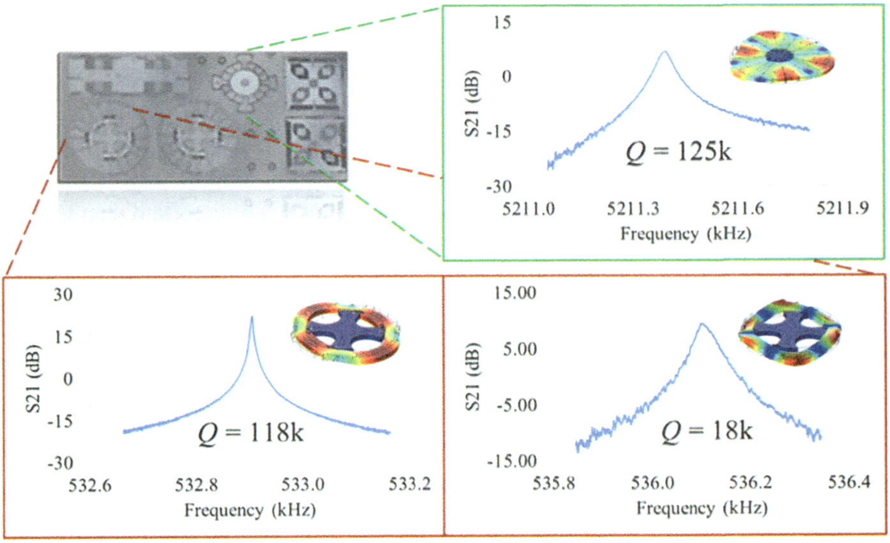

Fig. 3.21 Measured resonant peaks of a disk yaw gyroscope and the annulus pitch or roll gyroscope [9]

high-frequency gyroscopic modes with high Q-factors from both yaw and pitch/roll gyroscopes (Fig. 3.21).

3.4.3.2 Gyroscope Characterization

The annulus gyroscope is interfaced with an HF2LI lock-in amplifier for gyroscope operation testing. Electrical connection of the gyroscope is indicated in Fig. 3.22.

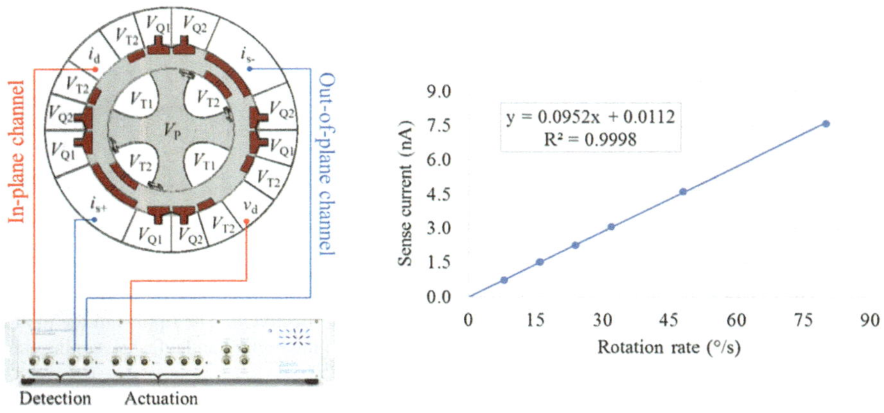

Fig. 3.22 Annulus gyroscope electrode configuration and measured mode-matched SF of 95.2 pA/(°/s) [9]

With the optimized tuning voltages, a mode-matched SF of 95.2pA/(°/s) is measured with an open-loop bandwidth of ~15 Hz (sense mode 3-dB bandwidth ~30 Hz).

The successful demonstration of mode-matched operation in the high-frequency pitch or roll annulus gyroscope with slanted electrodes is an important step toward high-performance MEMS IMUs, as it verified, for the first time, the possibility of implementing robust resonant pitch/roll gyroscopes with optimized output bias enabled by quadrature cancellation and substrate-decoupling. In addition, the HARPSS+ process shows its potential in multi-axis resonant gyroscope integration, providing a powerful tool for the implementation of more advanced high-frequency resonant gyroscope designs.

References

1. B.J. Gallacher et al., Electrostatic correction of structural imperfections present in a microring gyroscope. J. Microelectromech. Syst. **14**(2), 221–234 (2005)
2. D.E. Serrano, *Integrated Inertial Measurement Units Using Silicon Bulk-Acoustic Wave Gyroscopes* (Georgia Institute of Technology, 2014)
3. H. Wen, A. Daruwalla, F. Ayazi, Resonant pitch and roll silicon gyroscopes with sub-micron-gap slanted electrodes: Breaking the barrier toward high-performance monolithic inertial measurement units. Microsyst. Nanoeng. **3**, 16092 (2017)
4. S. Franssila, Anisotropic wet etching, in *Introduction to Microfabrication*, (John Wiley & Sons, Ltd, Hoboken, 2010), pp. 237–254
5. P. Prem, C. Sudhir, A novel process for perfect convex corner realization in bulk micromachining. J. Micromech. Microeng. **14**(10), 1416 (2004)
6. B. Kim et al., Temperature dependence of quality factor in MEMS resonators, in *19th IEEE International Conference on Micro Electro Mechanical Systems*, (2006), pp. 590–593

7. S. Ghaffari et al., Quantum limit of quality factor in silicon micro and nano mechanical resonators. Sci. Rep. **3**, 3244 (2013)
8. R. Tabrizian, M. Rais-Zadeh, F. Ayazi, Effect of phonon interactions on limiting the f.Q product of micromechanical resonators, in *TRANSDUCERS 2009–2009 International Solid-State Sensors, Actuators and Microsystems Conference*, (2009), pp. 2131–2134
9. H. Wen, A. Daruwalla, F. Ayazi, A 0.5MHz mode-matched pitch or roll annulus gyroscope with nano-gap slanted electrodes for quadrature cancellation, in *2017 19th International Conference on Solid-State Sensors, Actuators and Microsystems (TRANSDUCERS)*, (IEEE, 2017), pp. 536–539. © 2017 IEEE. Figures and tables reprinted with permission

Chapter 4
Scale-Factor Enhancement

A large SF is very important for MEMS gyroscopes to achieve small noise equivalent rotation rate and low rotation rate output bias. A smaller noise equivalent rotation rate directly translates to a smaller ARW and higher resolution. And a lower rotation rate output bias will lead to smaller rotation rate bias variations therefore lower bias instability. In this chapter, novel fabrication technique and gyroscope design will be presented to significantly enhance the SF and SNR of high-frequency MEMS resonant pitch or roll gyroscopes.

4.1 Transduction Linearity and Actuation Range

Efficient actuation of high-frequency gyroscopes is challenging due to their high stiffness. With HARPSS+ process, nano-gap parallel-plate capacitive transduction enables high electrostatic force density and low motional impedance in high-frequency resonators as shown in (2.36) and (2.39). However, the gap-changing mechanism is nonlinear by nature. As mentioned in Chap. 2, nonlinear transduction causes up-conversion of flicker noise to the sideband of the gyroscope drive-loop carrier frequency and adds flicker noise to the demodulated output [1]. Considering the first order nonlinearity in the electrostatic force:

$$F_{\text{elec}} \approx \frac{C_0 V^2}{2g_0}\left(1 + 2\frac{x}{g_0}\right) = \frac{C_0 (V_P + v_d + v_n)^2}{2g_0}\left(1 + 2\frac{x}{g_0}\right), \tag{4.1}$$

where V_P, v_d, v_n, C_0, g_0, and x are the polarization voltage, actuation voltage, noise voltage, the initial parallel-plate capacitance, initial gap-size, and the vibration displacement, respectively. For noise with low-frequency ω_n, the up-converted noise force is close to the resonant frequency. The linear drive force and noise force are:

H. Wen, *Toward Inertial-Navigation-on-Chip*, Springer Theses,
https://doi.org/10.1007/978-3-030-25470-4_4

$$\begin{aligned} F_{d@\omega_0} &= \frac{C_0 V_P v_d}{g_0} \\ F_{n@(\omega_0 \pm \omega_n)} &= \frac{C_0 v_d v_n}{g_0} + 2\frac{C_0 V_P v_n}{g_0^2}x \approx \frac{C_0 V_P v_d v_n}{g_0}\left(\frac{1}{V_P} - 2j\frac{QV_P C_0}{k_0 g_0^2}\right). \end{aligned} \tag{4.2}$$

The readout current SNR is then:

$$i_d/i_n \approx \left[\left(\frac{1}{V_P} - 2j\frac{QV_P C_0}{k_0 g_0^2}\right)v_n\right]^{-1}. \tag{4.3}$$

As (4.3) shows, the low-frequency SNR due to transduction nonlinearity is independent of drive amplitude. Therefore, increasing drive amplitude will not help to suppress the effect of flickers noise in drive-loop.

At the same time, for large displacements, higher-order nonlinearity needs to be considered and (4.1) becomes:

$$\begin{aligned} F_{\text{elec}} &= \frac{V^2}{2}\frac{\partial}{\partial x}\left(\frac{\varepsilon A}{g_0 - x}\right) = \frac{\varepsilon A V^2}{2(g_0 - x)^2} \approx \frac{C_0 V^2}{2g_0}\left[1 + 2\frac{x}{g_0} + 3\left(\frac{x}{g_0}\right)^2 + 4\left(\frac{x}{g_0}\right)^3\right] \\ &= \frac{C_0 V^2}{2g_0} - x\left(-\frac{C_0 V^2}{g_0^2} - \frac{3C_0 V^2}{2g_0^3}x - \frac{2C_0 V^2}{g_0^4}x^2\right) \\ &= \frac{C_0 V^2}{2g_0} - x\left(-\frac{C_0 V^2}{g_0^2} - \frac{3C_0 V^2}{2g_0^3}x - \frac{2C_0 V^2}{g_0^4}x^2\right) \end{aligned} \tag{4.4}$$

As shown by (4.4), parallel-plates not only provide a linear actuation force but also add nonlinear spring constants to the gyroscope system. With a cubic nonlinear restoring force, the drive-loop becomes a Duffing oscillator exhibiting an amplitude-frequency (*A-f*) effect. The resonant peak shifts away from the natural frequency as the vibration amplitude increases. The shifted frequency is given by:

$$f = (1 + \kappa x^2)f_0 \text{ and } \kappa = \frac{3k_{e2}}{8k_0} - \frac{5k_{e1}^2}{12k_0^2}, \tag{4.5}$$

where κ is called the *A-f* coefficient, and k_0 is the sum of the mechanical and electrostatic linear stiffness of the oscillator. Since k_{e2} is always negative, the *A-f* effect will cause the peak to bend to a lower frequency, which is called nonlinear spring softening effect. After a certain amplitude, the amplitude-frequency curve is no longer a single-valued function. And the largest vibration amplitude before bifurcation is called the critical amplitude x_c [2] (Fig. 4.1). Large *A-f* effect will cause frequency and phase errors in presence of oscillation amplitude fluctuation, therefore causes additional bias instability as described in (2.63). Therefore, the drive displacement amplitude is usually kept to less than 10% of the gap-size, which

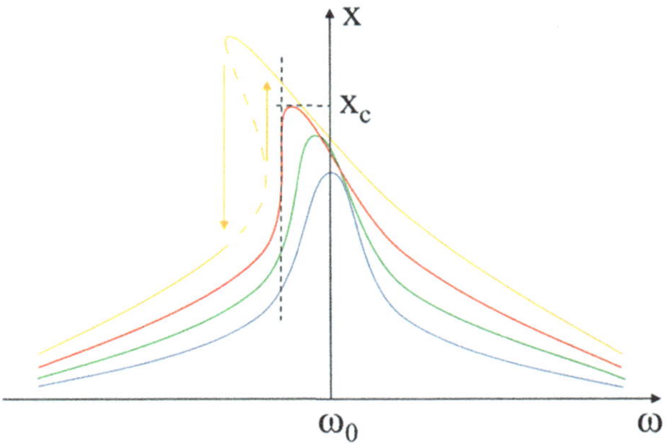

Fig. 4.1 Frequency response of a Duffing oscillator with a spring softening effect under different actuations level

is referred to as the linear rive range. For a typical HARPSS+ gap of 200 nm, the linear drive range will only be 20 nm, which will lead to a higher thermal noise equivalent rate output as described in (2.57).

Therefore, it is very desirable to improve the transduction linearity and extend the linear actuation range. Comb-drives utilizing transduction area changing mechanism are linear transducers by nature, and they are widely used in low-frequency MEMS sensors. However, due to fabrication limits, comb-drives are usually implemented with interdigitated comb-fingers with capacitive gaps in the range of a few microns, yielding insufficient force to actuate high-frequency resonators or gyroscopes, or having large form factors [3]. How to build low nonlinearity capacitive transducers with small size and efficient nano-gap transduction becomes very important to enable both low ARW and low BI in high-frequency gyroscopes.

4.1.1 Nano-gap Comb-Drive Design

Conventional comb-drives rely on the change of overlapping area between parallel-plates from two set of comb-fingers with continuous surfaces. In HARPSS+ process, thermal oxide is used to define nano-gaps between SCS and polysilicon comb-fingers. Polysilicon fingers are formed by selective DRIE of polysilicon from poly-filled trenches. Their sidewall profile depends on lithography and etching accuracy. The sidewall surface is continuous only in an ideal case with perfect alignment and etching. In reality, there will always be lithography misalignment errors and etching imperfections. With the gap-size being in nanoscale, common misalignment error can easily cause surface discontinuity and steps that are comparable or even larger than the gap-size, therefore preventing the proper function of the comb-drive (Fig. 4.2).

To work around the surface continuity requirement, comb-drives with gap-closing comb-fingers may be used where the fingers are slightly tapered and

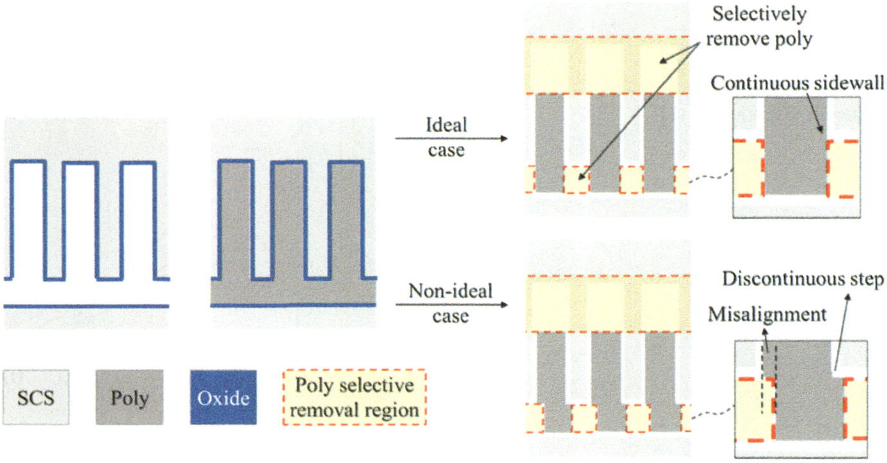

Fig. 4.2 Steps for implementation of conventional comb-drives in HARPSS+ process: (1) SCS DRIE to form one set of comb-fingers and thermal oxidation to define gap-size; (2) polysilicon trench filling; (3) selective poly etching to form the other set of comb-fingers [5]

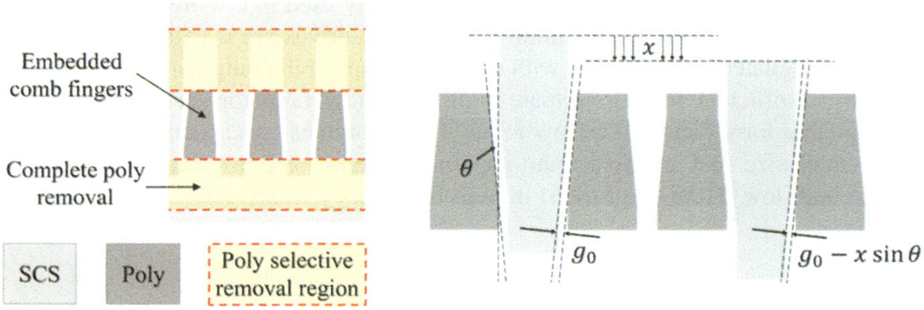

Fig. 4.3 Top view of tapered comb-fingers fabricated with HARPSS+ process and schematics illustrating the operation principle [5]

the gap-size varies when the two sets of fingers move relatively [4, 5]. With gap-closing operation, the change of overlapping area is no longer a requirement. Therefore, one set of the fingers can be shorter than the other set and completely embedded in the spaces between the other set of fingers. This way, polysilicon can be etched off completely on both ends of the poly fingers and then be connected together through bridges on top. The etching process involved in removing polysilicon from both ends is not sensitive to misalignment errors, and small lithography and etching errors will not significantly affect the functionality of the fingers (Fig. 4.3). By going to the gap-closing operation, the comb-drive will have a finite linear range. However, with a small tapering angle, the linear drive range can be sufficient for most high-frequency gyroscope applications. For a tapering angle θ, the electrostatic force and the linear drive range of the comb-drive are given by:

$$F_{\text{elec}} = \frac{V^2}{2}\frac{\partial}{\partial x}\left(\frac{2n\varepsilon A}{g_0 - x\sin\theta}\right) \approx \sin\theta\frac{C_0V^2}{2g_0}\left(1 + 2\frac{\sin\theta x}{g_0}\right) \tag{4.6}$$

$$x_{\text{linear}} = 10\% \times \frac{g_0}{\sin\theta}, \tag{4.7}$$

where n and A are the number of fingers in each set and the transduction area between each pair of fingers. It can be seen from (4.6) and (4.7) that using small θ provides large linear drive range with the cost of lowering the drive force. However, the reduction in electrostatic force can be compensated by increasing the initial capacitance with a larger number of fingers. Comparing (4.1) and (4.6) shows, to achieve a similar level of electrostatic force as the nano-gap parallel-plate, we should have:

$$C_{0\text{comb}}\sin\theta \approx C_{0\text{parallel}} \tag{4.8}$$

Following (4.1), (4.2), and (4.3), the readout current SNR with tapered comb-drive is:

$$\begin{aligned} \frac{i_d}{i_n} &\approx \left[\left(\frac{1}{V_P} - 2j\frac{QV_PC_{0\text{comb}}}{k_0g_0^2}\sin^2\theta\right)v_n\right]^{-1} \\ &\approx \left[\left(\frac{1}{V_P} - 2j\frac{QV_PC_{0\text{parallel}}}{k_0g_0^2}\sin\theta\right)v_n\right]^{-1}. \end{aligned} \tag{4.9}$$

For high-Q resonator with nano-gap transduction, the second term is generally dominant. Therefore (4.3) and (4.9) give:

$$\left(\frac{i_d}{i_n}\right)_{\text{comb}} \approx \left(\frac{i_d}{i_n}\right)_{\text{parallel}}\sin\theta. \tag{4.10}$$

Equations (4.7) and (4.10) show that the tapered nano-gap comb-drive can provide improvements in both linear actuation range and flicker noise rejection when generating the same level of actuation force as a nano-gap parallel-plate actuator.

4.1.2 Testing Resonator Design

To verify the functionality of the nano-gap comb-drive design, a testing high-frequency annulus resonator is used with the nano-gap comb-drive as the driving

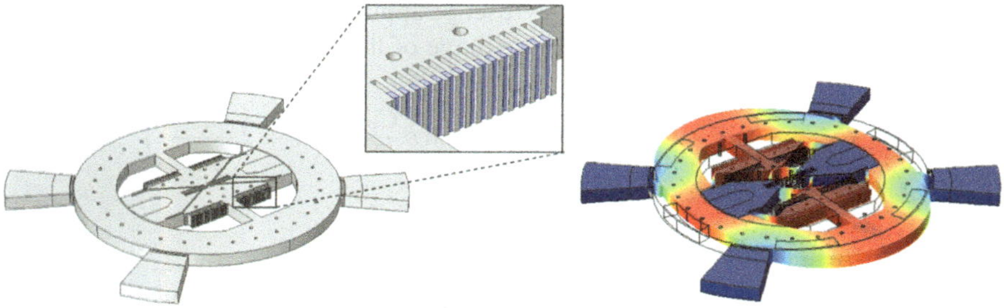

Fig. 4.4 Resonant structure and the comb-driven $n = 2$ wineglass mode of the high-frequency annulus resonator. Inset shows the SCS (gray) and polysilicon (blue) comb-fingers (top connection of the polysilicon fingers not shown) [5]

mechanism for an $n = 2$ wineglass mode at 1 MHz. To better examine the linearity of the nano-gap comb-drive by observing the *A-f* phenomenon, effect mechanical nonlinearity of the testing resonator should also be considered. Unlike electrical nonlinearity, mechanical nonlinearity can cause either spring softening or spring hardening effect. Having a spring softening mechanical nonlinearity may prevent the accurate determination of electrical nonlinearity level; therefore an annulus resonator with spring hardening mechanical nonlinearity is preferred.

Mechanical nonlinearity of a SCS resonator depends on its mode-shape, doping level, and crystal orientation. Available studies on the mechanical nonlinearity of $n = 2$ wineglass mode of SCS annulus resonators are very limited due to the special geometry and complicated strain distribution of the mode. To obtain an estimation of the mechanical nonlinearity of the high-frequency $n = 2$ mode of annulus implemented on different types of substrate, an analogy is built between the annulus and a length-extension resonator with scaled Young's modulus, which can be modeled using a lumped-element approximation. Numerical simulations using the lumped-element model indicate that annulus resonator with spring hardening nonlinearity can be achieved on N-doped (100) substrate with the $n = 2$ mode aligned to (100) direction. Therefore, a design is developed accordingly for the nano-gap comb-drive testing (Fig. 4.4). More details on the numerical estimation of the annulus mechanical nonlinearity are listed in Appendix B.1.

4.1.3 Experimental Characterization

The comb-driven annulus resonator design is implemented on a (100) SOI wafer with 40-μm-thick, 15-mΩ·cm phosphorus-doped device layer, with the comb-drive designed to provide a linear drive range larger than 400 nm and a force density ~250 $pN/V^2\mu m$ with a gap-size of 190 nm. Figure 4.5 shows the SEM pictures of a fabricated device.

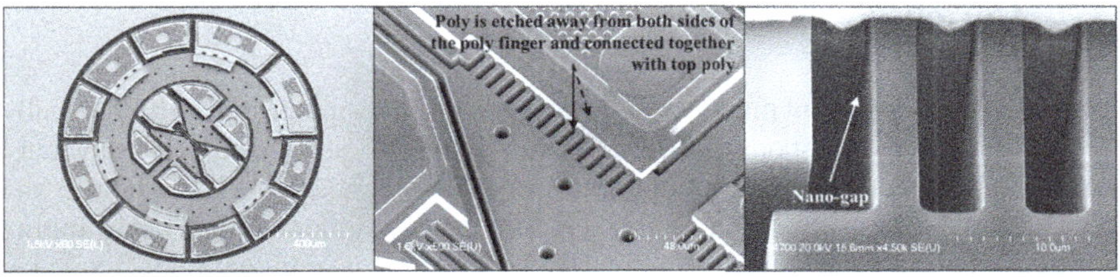

Fig. 4.5 SEM picture of fabricated comb-drive annulus BAW resonator and zoom-in pictures of the comb-drive structure. The comb-fingers are separated by 190 nm capacitive gaps [5]

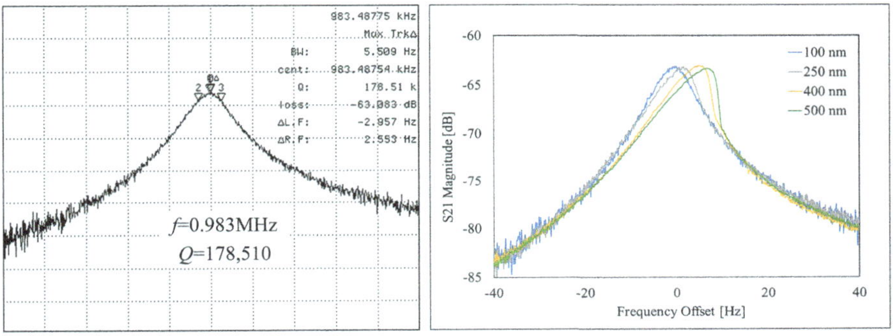

Fig. 4.6 Resonant peak and nonlinearity characterization of the nano-gap comb-driven annulus resonator [5]

Characterizing the fabricated resonator with a network analyzer in a vacuum probe station at 1 m Torr shows a high Q of ~178,000 for the $n = 2$ wine-glass mode at 0.983 MHz. The force density of the comb-drive is estimated to be ~260 pN/V^{2}μm based on the measured motional impedance, which is in good agreement with the theoretical value and is much larger than conventional silicon MEMS comb-drive actuators.

Nonlinearity of the nano-gap comb-driven resonator is characterized at different drive amplitudes by increasing v_d with constant V_P. Critical amplitude larger than 400 nm is measured. As the drive amplitude increases, no spring softening effect occurred. For drive amplitude larger than 200 nm (10× larger than parallel-plate linear actuation range with the same gap-size), spring hardening effect starts to be observed, which agrees with the design expectation and verifies the reduced electrical nonlinearity of the comb-drive actuator (Fig. 4.6).

Characterizing the nano-gap comb-drive with the testing resonator verifies the efficiency of the actuator for providing large electrostatic actuation force with a small form factor as well as the reduction of nonlinear effect and extended linear actuation range. The nano-gap comb-drive concept can be modified and adapted in various high-frequency gyroscopes design, extending the drive amplitude while reducing nonlinear effects, therefore improving their SF and SNR.

4.2 Coriolis Sensitivity

High-frequency resonant pitch or roll gyroscopes take advantage of the mechanically Q-amplified sense response, which allows them to use small-size substrate-decoupled resonant structures with better robustness against shock, vibration, and temperature variations compared to their low-frequency counterpart using tuning fork designs and mode-split operation. However, the conventional high-frequency annulus-based designs presented so far suffer from low gyroscopic coupling between the drive and sense modes (i.e., Coriolis sensitivity) and consequently low SF and SNR. Efforts have been put into modifying the annulus dimensions such as changing annulus width distribution and adding ovalness to improve the Coriolis sensitivity, but effects have been marginal. Therefore, further improving the performance of high-frequency resonant pitch or roll gyroscopes for navigational applications calls for a breakthrough in the design concept to construct novel structures with high Coriolis coupling modes as well as good substrate isolation and environmental robustness.

4.2.1 Mode-Shape Optimization

The Coriolis coupling is essentially what separates the rotation rate responds from noise-induced output. A higher Coriolis sensitivity means the gyroscope generates larger response to rotation input than noise in the system. From (2.35) and (2.57), we can see the SNRs for both white noise and flicker noise components are proportional to the Coriolis-coupling coefficient. Therefore, improving Coriolis sensitivity will lead to improvements in both ARW and BI.

As defined in Chap. 2, the Coriolis-coupling coefficient is a mode-shape-dependent constant with value between 0 and 1. Denoting $U_{1,\,2}$, $U_{x1,\,2}$, $U_{y1,\,2}$, and $U_{z1,\,2}$ to be mode-shape functions for the normalized displacement, the x-component, the y-component, and the z-component of normalized displacement field of drive and sense modes, respectively. Following (2.5), the Coriolis-coupling coefficient for a roll gyroscope is:

$$\lambda = \frac{\oint \left| U_{y1} U_{z2} - U_{z1} U_{y2} \right| dV}{\oint U_2^2 dV}. \tag{4.11}$$

A Coriolis-coupling coefficient of 1 means the Coriolis force distribution generated by drive mode displacement under rotation matches the sense mode deformation distribution perfectly. For a roll gyroscope, this means:

$$U_{x1} = 0, \quad U_{y1} = U_{z2}, \quad \text{and} \quad U_{z1} = -U_{y2}. \tag{4.12}$$

On the other hand, a Coriolis-coupling coefficient less than 1 means part of the sense mode effective mass is not directly actuated by the Coriolis force due to any deviation from the ideal mode-shape conditions given in (4.12). In such cases, the total energy transferred from drive mode to sense mode under rotation through the Coriolis coupled mass will partially be consumed to actuate the uncoupled mass, resulting in a reduced sense mode excitation level overall. Equation (4.12) therefore provides the guidelines for constructing resonant structures with high Coriolis sensitivity.

The other important consideration when designing mode-shapes for robust Coriolis gyroscopes is the self-balancing and substrate independence of the mode motions. In order to achieve high- robustness substrate-decoupled operation modes, the mode motions should satisfy the conservation of momentum and angular momentum so that no external balancing force is required from the substrate the sustain the vibration. The mathematical expression of the self-balanced conditions is:

$$\oint \vec{U}_1 dV = \oint \vec{U}_2 dV = 0 \text{ and} \oint \vec{U}_1 \times \vec{r} dV = \oint \vec{U}_2 \times \vec{r} dV = 0, \tag{4.13}$$

where $\vec{r}$ is the position vector of an infinitesimal element relative to fixed reference point in the gyroscope local frame. A pair of optimized gyroscopic modes should satisfy or close to satisfy both (4.12) and (4.13) for best performance delivery.

Two examples can be examined to better understand the two sets of mode-shape requirements. First, the traditional tuning fork roll gyroscope design uses an IP mode with mirror-symmetric motions of two tines (proofmass) along y-axis as the drive mode and an OOP mode with mirror-asymmetric motions of the two tines along z-axis as sense mode, or vice versa (Fig. 4.7). The normalized mode-shape functions of the two modes are approximately given by:

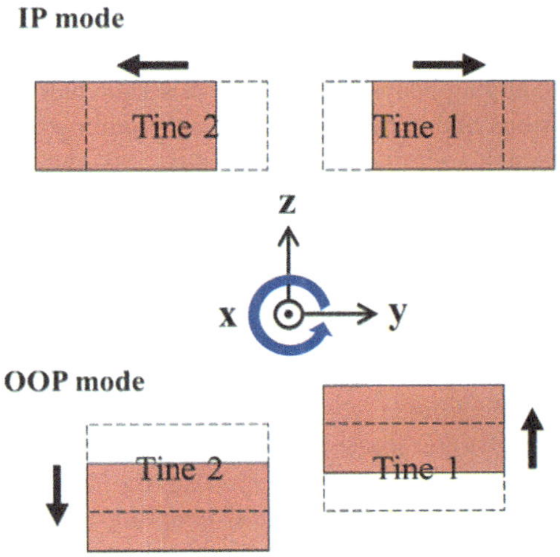

Fig. 4.7 Schematic illustration of the proofmass motions for the drive and sense modes in a tuning fork roll gyroscope [7]

$$U_1 = U_{y1} = \begin{cases} 1 & @tine\ 1 \\ -1 & @tine\ 2 \end{cases} \text{ and } U_2 = U_{z2} = \begin{cases} 1 & @tine\ 1 \\ -1 & @tine\ 2 \end{cases}. \tag{4.14}$$

The mode-shape function (4.14) satisfies (4.12) but not (4.13). Physically, this means tuning fork pitch or roll gyroscopes provide optimized Coriolis sensitivity, but the asymmetric OOP mode has an unbalanced angular momentum, requiring additional balancing force from the anchors and substrate. The coupling to the substrate not only will increase the anchor loss of the OOP mode but also makes it sensitive to changes in the substrate like stress induced by chip attachment and ambient temperature variations.

Second, the conventional high-frequency annulus pitch or roll gyroscope uses IP $n = 2$ wineglass mode as drive mode and OOP $n = 3$ bending mode as the sense mode, or vice versa. The operational modes have mode-shape functions with second and third angular periodic displacement distributions. Therefore, volume integration of momentum and angular momentum of both modes are 0, meeting condition (4.13) and allowing the modes to be decoupled from the substrate while anchored rigidly with low anchor loss. However, condition (4.12) is violated because the IP mode y-component displacement is first-order angular periodic, but the OOP mode z-component displacement is third-order angular periodic. Analytical expressions of the mode-shape functions based on combinations of Bessel and trigonometric functions are available for the IP and OOP modes and the Coriolis-coupling coefficient of the annulus can be calculated using (4.11). Alternatively, the mode-shapes can be evaluated numerically using FEM simulations. The simulation results for an ideal annulus shows a Coriolis-coupling coefficient of 0.25. In practice, inevitable deviations from the ideal annulus structure such as additional tethers may further reduce the Coriolis-coupling coefficient to around 0.2 (Fig. 4.8).

To create a pitch or roll gyroscope meeting both criteria, a novel framed-annulus gyroscope is developed [7]. The new design uses a bulk annulus structure as a starting point. To create self-balanced OOP motion without using a higher-order periodic mode-shape, an outer frame is introduced to enable extra design freedoms for mode-shape engineering, which is coupled to the inner annulus through rigid coupling beams. A pair of gyroscopic modes is constructed by adjusting the dimensions of the frame, annulus, and coupling beams. For the IP mode, the frame and the annulus vibrate in an antiphase (180° out-of-phase) $n = 2$ wine-glass deformation. Like a regular $n = 2$ wine-glass mode, the antiphase $n = 2$ wine-glass mode is second-order angular periodic and therefore satisfies condition (4.13). For the OOP mode, the frame and the annulus vibrate with an antiphase teeter-totter motion. At the antinodes, the annulus, and the frame move in opposite directions, counterbalancing each other and enabling the conservation of momentum and angular momentum, therefore also satisfies (4.13). Unlike the third-order OOP bending mode which has third-order angular periodic z displacements, the antiphase teeter-totter mode has first-order angular periodic z displacements, which match the y displacement distribution in the IP mode therefore maximizing the Coriolis coupling with the Coriolis-coupling coefficient close to 1. The balanced motions

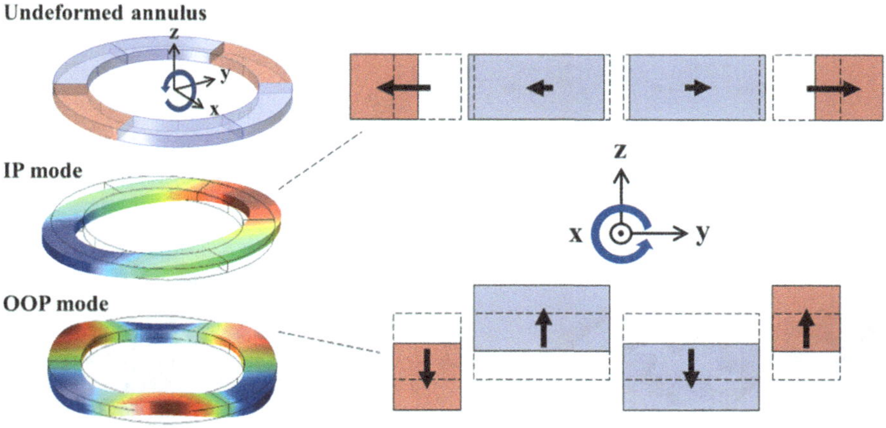

Fig. 4.8 Geometry and mode-shapes of an ideal annulus roll gyroscope. The undeformed annulus is color-labeled to indicate Coriolis coupled and uncoupled regions. Mode-shapes are shown with color distribution showing the y displacement and z displacement for the IP and OOP modes, respectively [7]

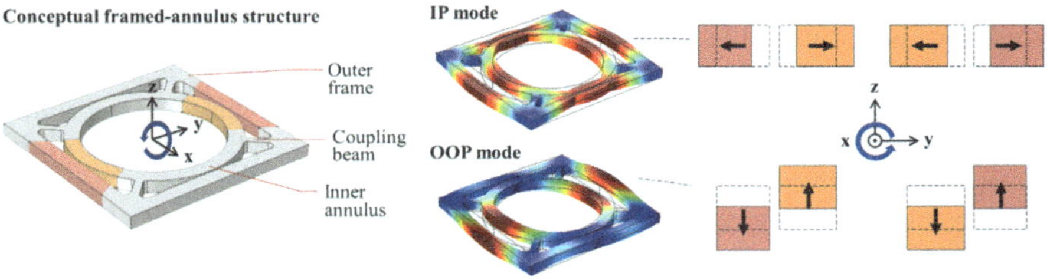

Fig. 4.9 Conceptual framed-annulus structure and the substrate-independent IP and OOP mode-shapes [7]

of the frame and the annulus also create common nodes for the IP and OOP operational modes near the corners of the frame, providing ideal locations for substrate-decoupled low-loss anchoring (Fig. 4.9).

4.2.2 *Resonant Framed-Annulus Pitch or Roll Gyroscope*

A prototype resonant pitch or roll gyroscope design is developed implementing the framed-annulus concept with an operational frequency of ~660 kHz. Figure 4.10 shows the resonant structure of the prototype and its operational mode-shapes. The IP mode is used as the drive mode, in which a modified version of the nano-gap comb-drive introduced in the previous section is used to extend the linear actuation range. Top electrodes are used for sense readout and frequency tuning. Frequency

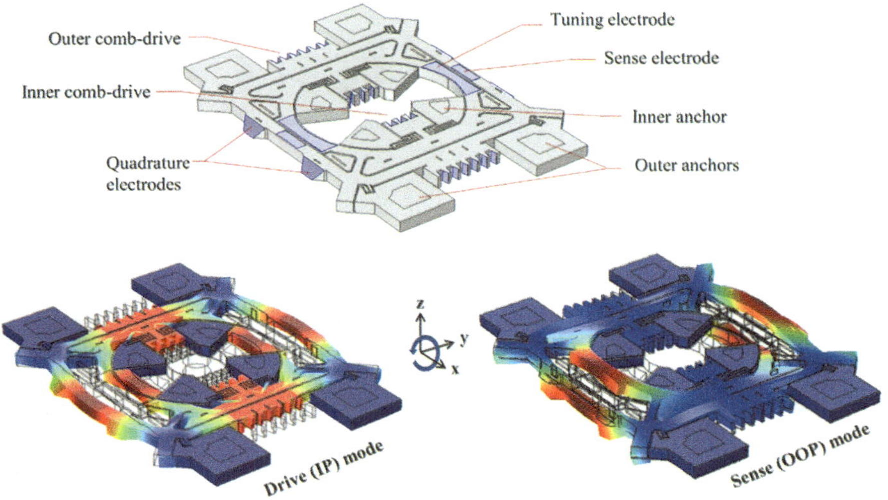

Fig. 4.10 Framed-annulus resonant roll gyroscope prototype resonant structure and operational mode-shapes. Capacitive transduction regions are labeled in blue on the resonant structure [7]

difference between the IP and OOP modes is designed by adjusting the geometry of the coupling beams to close-to-mode-matched condition, with the untuned OOP frequency slightly higher than the IP mode to ensure mode-matching capability through electrostatic softening. Slanted electrodes are integrated on the edge of the outer frame for quadrature tuning as discussed in Chap. 3.

Short rigid tethers are added near the common nodes of the IP and OOP modes with low anchor loss as mentioned earlier. In addition, serpentine tethers are attached to the inner comb-drives to enhance the OOP stability. FEM simulations with perfectly matched layer [6] show the overall tether structure provides a substrate-decoupled anchoring with Q_{anchor} above 6.5 million for both modes across process variations, much higher than the simulated Q_{TED} of less than 60k. The rigid tethers also lead to high translational stiffness of the resonant structure. Simulation shows all resonance modes of the structure are above 200 kHz, higher than the spectrum of environmental random vibrations (typically below 50 kHz) (Figs. 4.11 and 4.12). The Coriolis-coupling coefficient of the gyroscope is simulated to be 0.83, which is ~4× higher than the conventional annulus pitch or roll gyroscopes and also higher than BAW disk yaw gyroscopes using $n = 3$ degenerate modes (λ~0.6).

4.2.3 Experimental Characterization

The prototype is fabricated using the HARPPS+ process on a (100) SOI base wafer with 40um-thick device layer. Electrode gap-size of ~270 nm is chosen to provide sufficient tuning coverage for resonant operation. The completed based wafer is wafer-level packaged using vacuum eutectic bonding to a capping wafer with built-

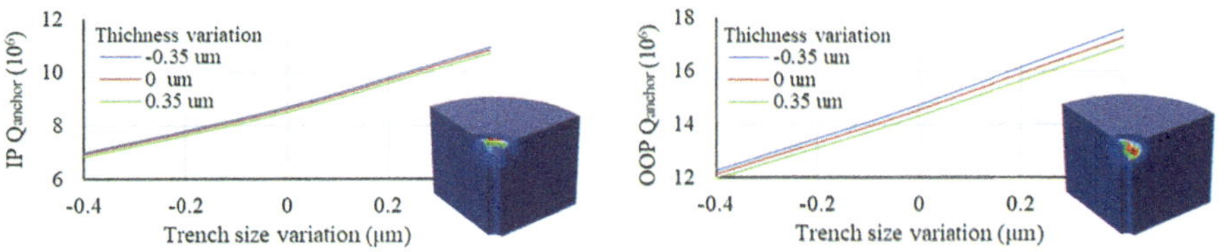

Fig. 4.11 Simulated anchor loss of the IP and OOP operational modes across fabrication variations. Insets show the PML deformations for each mode

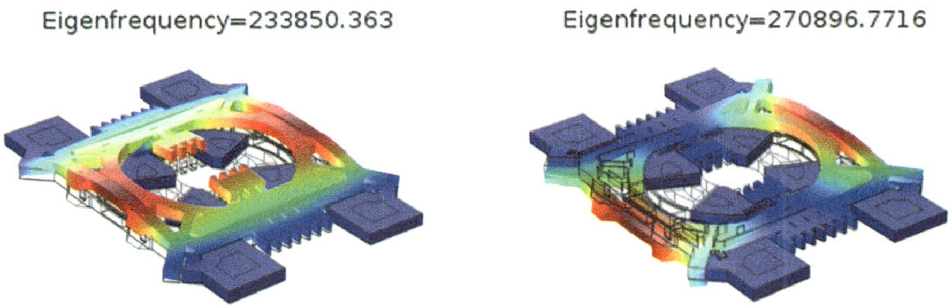

Fig. 4.12 Simulated mode-shapes of the two modes with the lowest resonant frequencies [7]

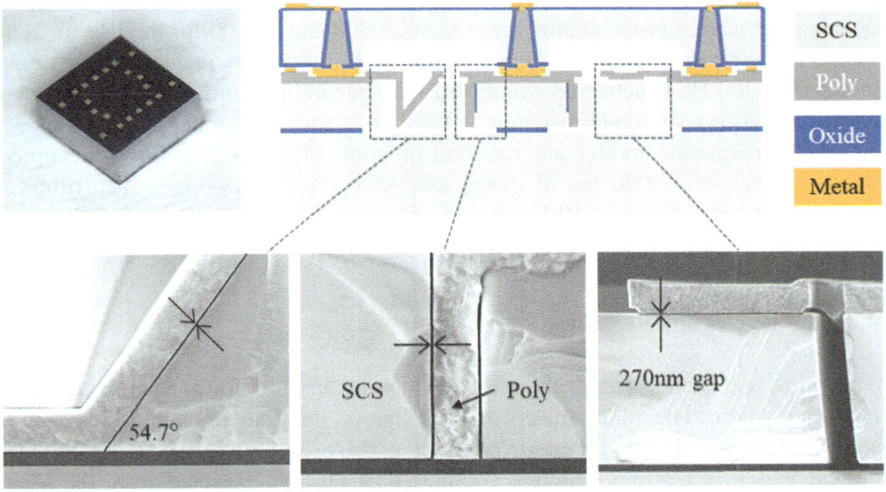

Fig. 4.13 WLP pitch or roll gyroscope chip fabricated using HARPSS+ process and the typical cross-sectional views of different electrodes [7]

in cavities and through silicon vias. The resonant structure of the device is a 0.9 mm × 0.9 mm, and the WLP die is 2 mm × 2 mm (Fig. 4.13). Figure 4.14 shows the SEM images of the fabricated framed-annulus gyroscope. The fabricated framed-annulus gyroscopes are wire-bonded to a printed circuit board and characterized with discrete interface electronics.

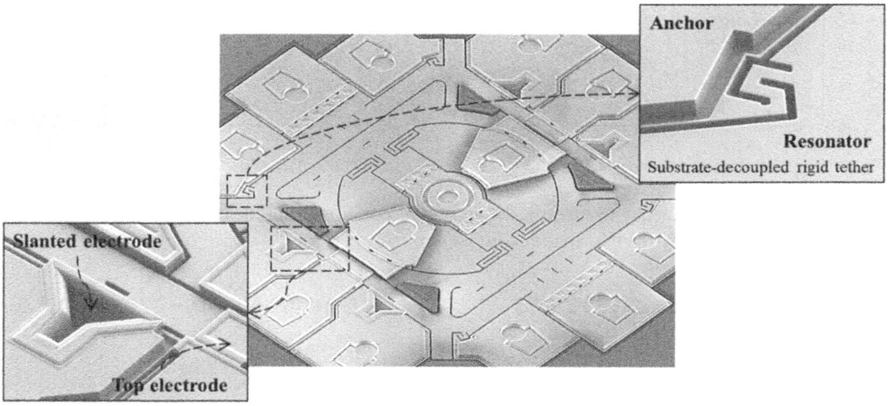

Fig. 4.14 SEM image of fabricated framed-annulus pitch or roll gyroscope. Insets show zoomed-in views of the electrodes and the rigid substrate-decoupled tether design [7]

4.2.3.1 Resonant Behavior

The WLP gyroscope is first measured using a 4-port network analyzer. Measurements show Q of 13,000 and 4500 due to SFD for the IP and OOP modes, respectively, indicating a package pressure of ~5.5 Torr. Measuring an uncapped device in a vacuum probe-station gives Q of 29,000 and 51,000 for the IP and OOP mode, respectively, matching the simulated TED limits. Frequency tuning range larger than 20 kHz is achieved, combined with quadrature tuning using the slanted electrodes, allows the mode-matching at V_P of 16 V and with proper tuning voltages. The two operational modes are matched at ~660 kHz showing fully decoupled quality-factors of 13,000 for IP mode, and 4500 for OOP mode, which offers a large 3 dB bandwidth of ~150 Hz (Fig. 4.15).

4.2.3.2 Gyroscope Response

The gyroscope is interfaced with an HF2LI lock-in amplifier for gyroscope response characterization. The mode-matched operation of the framed-annulus gyroscope shows a current SF of 150 pA/(°/s) with a drive amplitude of 80 nm, which corresponds to a Coriolis-coupling coefficient of ~0.8 as expected from the FEM simulations. Figure 4.16 shows the SF measurement plot as well as the stable time-domain response of the gyroscope under rotations with different amplitudes and frequencies. The full-scale range of the gyroscope is verified to be larger than 600 °/s (limited by testing setup capability) with a full-scale nonlinearity smaller than 0.08%.

Figure 4.17 shows the measured ADEV of the resonant framed-annulus gyroscope. Measurement shows an ARW of 0.55 °/√h and a BI of 17.9 °/h, as well as a

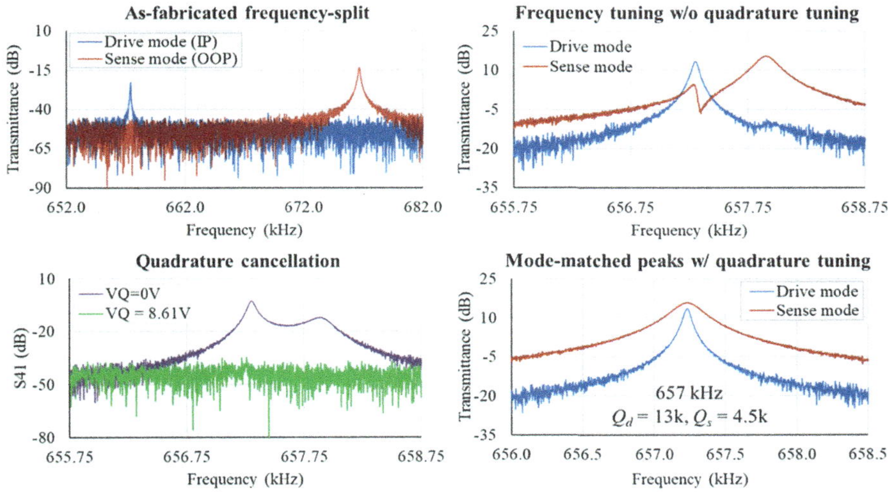

Fig. 4.15 Typical electrostatic tuning and mode-matching behaviors of the framed-annulus pitch or roll gyroscope [7]

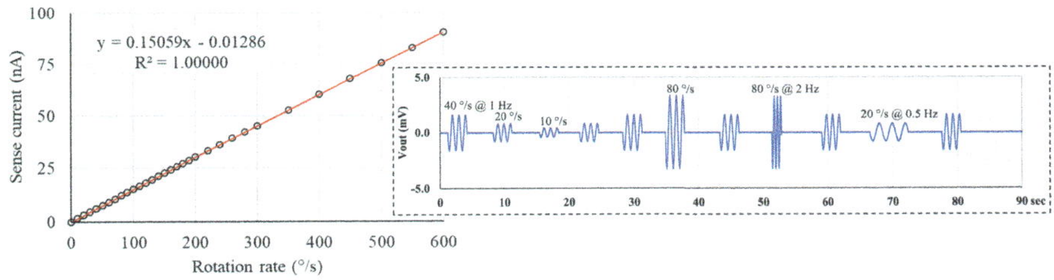

Fig. 4.16 SF measurement results and typical time-domain response of the resonant framed-annulus pitch or roll gyroscope [7]

Fig. 4.17 ADEV measurement result of the resonant framed-annulus pitch or roll gyroscope [7]

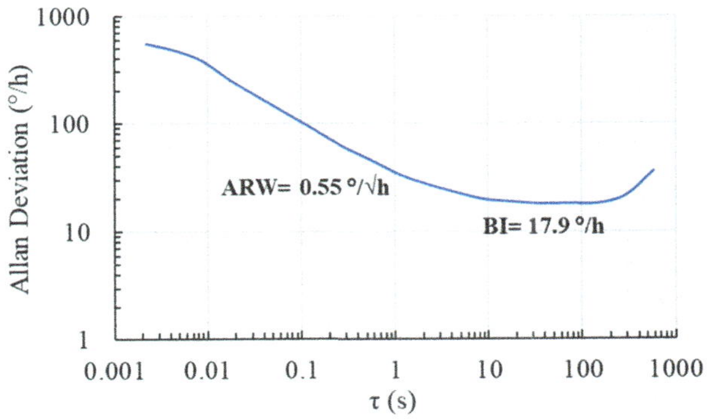

good long-term stability with no rate random walk (RRW) until averaging time larger than 5 minutes. These performance specs are by far the best ever reported for a high-frequency resonant pitch or roll gyroscope. Improving the package pressure will further reduce the ARW due to less SFD and higher SF. With a package pressure below 1 Torr, the ARW will be improved toward a TED-defined mechanical limit of 0.1 °/√h.

4.2.3.3 Temperature Behavior

The quality-factors of the IP and OOP operational modes are recorded at temperatures from −40 to 80 °C inside a temperature chamber. A well-behaved temperature dependence is found, which verifies the elimination of erratic changes due to anchor loss with the substrate-decoupled design. The SF of the framed-annulus gyroscope is also characterized from −40 °C to 80 °C. Frequency tuning voltage V_T is adjusted following a linear slope of 0.9 mV/°C to maintain mode-matched operation at different temperatures. The changes in SF agree with the changes in sense mode Q, giving a SF variation of less than ±28% across the entire temperature range. However, since the temperature dependence of Q and SF is monotonic and predictable, a quadratic compensation curve can be used to reduce the SF variation to less than ±0.65% (Fig. 4.18). Table 4.1 summarizes the measured performance of the resonant framed-annulus pitch or roll gyroscope.

The novel framed-annulus design has high-frequency operational mode-shapes optimized for (4.12) and (4.13) to achieve substrate-decoupling and high Coriolis sensitivity at the same time. When combined with the other performance optimization technologies developed in this work including the slanted quadrature tuning electrode and nano-gap comb-drive, the framed-annulus gyroscope demonstrates a high performance among the best of all MEMS pitch or roll gyroscopes with a smaller, more robust structure and a higher operational frequency. Even higher performance is expected from the same design with improvements in the package pressure and interface electronics. Fabricated using the advanced HARPSS+

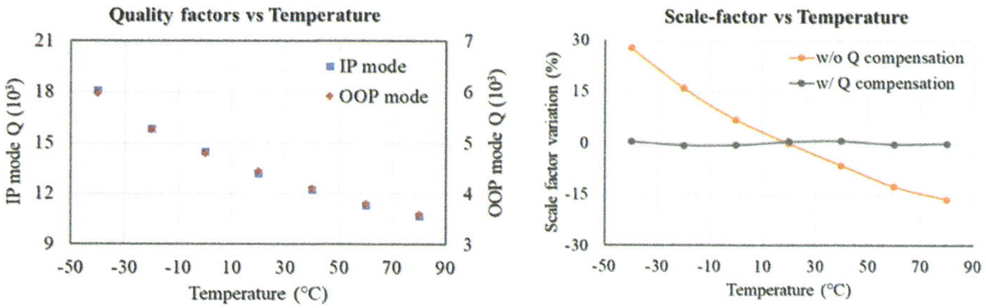

Fig. 4.18 Temperature behavior of the quality-factor and the scale-factor of the resonant framed-annulus pitch or roll gyroscope [7]

Table 4.1 Performance summary for the resonant framed-annulus gyroscope [7]

Parameter	Value	Unit
Resonator size	0.9 × 0.9	mm × mm
Operational frequency	657	kHz
Lowest resonant frequency	230	kHz
Mode-matched *Q*	13k (IP); 4.5k (OOP)	
Open-loop bandwidth	73	Hz
Coriolis-coupling coefficient	0.8	
Scale-factor	150	pA/(°/s)
Full-scale range	600	°/s
SF nonlinearity	0.08	%
Compensated SF temperature stability	0.65	%
ARW	0.55	°/√h
Bias instability	17.9	°/h
TED limited *Q*	29k (IP); 51k (OOP)	
Mechanical ARW limit due to TED	0.1	°/√h

process, the resonant framed-annulus pitch or roll gyroscope is integration-ready with high-frequency BAW resonant yaw gyroscopes to realize robust high-performance triaxial gyroscope units and 6-DOF IMUs.

References

1. V. Kaajakari, J.K. Koskinen, T. Mattila, Phase noise in capacitively coupled micromechanical oscillators. IEEE Trans. Ultrason. Ferroelectr. Freq. Control **52**(12), 2322–2331 (2005)
2. V. Kaajakari et al., Nonlinear limits for single-crystal silicon microresonators. J. Microelectromech. Syst. **13**(5), 715–724 (2004)
3. T. Hirano et al., Design, fabrication, and operation of submicron gap comb-drive microactuators. J. Microelectromech. Syst. **1**(1), 52–59 (1992)
4. M.A. Rosa, S. Dimitrijev, H.B. Harrison, Enhanced electrostatic force generation capability of angled comb finger design used in electrostatic comb-drive actuators. Electron. Lett. **34**(18), 1787–1788 (1998)
5. H. Wen, et al., HARPSS-fabricated nano-gap comb-drive for efficient linear actuation of high frequency BAW resonators, in *2016 IEEE 29th International Conference on Micro Electro Mechanical Systems (MEMS)*, (IEEE, 2016), pp. 1014–1017. © 2016 IEEE. Figures and tables reprinted with permission
6. D.S. Bindel, S. Govindjee, Elastic PMLs for resonator anchor loss simulation. Int. J. Numer. Methods Eng. **64**(6), 789–818 (2005)
7. H. Wen, A. Daruwalla, C.S. Liu, F. Ayazi, A high-frequency resonant framed-annulus pitch or roll gyroscope for robust high-performance single-chip inertial measurement units. *Journal of Microelectromechanical Systems* **27**(6), 995–1008 (2018). © 2018 IEEE. Figures and tables reprinted with permission

Chapter 5
Integrated Inertial Measurement Unit

With the more advanced pitch and roll gyroscope designs, achieving consistent high performance in all three axes using robust high-frequency resonant MEMS gyroscopes becomes possible. At the same time, the HARPSS+ process provides an ideal platform to implement and integrate the multi-DOF high-performance designs. In this chapter, an integrated TIMU will be presented with great potentials demonstrated from each of the integrated sensing elements [1].

5.1 Quasi-Solid Disk Yaw Gyroscope

A quasi-solid BAW disk gyroscope is developed for robust yaw rotation sensing in the IMU. The gyroscope uses $n = 3$ degenerate wine-glass modes as gyroscopic modes. Like in pitch and roll gyroscopes, energy localization of the modes is also important in yaw gyroscopes for achieving stable high Q and isolating the acoustic wave from disturbances in the substrate. This can be done using flexures with high degrees of freedom to virtually levitate the device. However, inclusion of a decoupling beam network is accompanied by increased TED and reduced translational frequencies. Alternatively, substrate decoupling can be achieved using carefully designed notches to shield the acoustic wave from non-idealities like material anisotropy [2]. In the quasi-solid disk gyro, we took this approach with an improved narrow decoupling slots design, which enables a mode-specific substrate-isolated structure with a form closest to a solid disk therefore possessing some of its advantages. Particularly, like a solid disk, the quasi-solid disk is extremely stiff in translational directions with all its translational modes above 1 MHz. As a result, it provides superior robustness against environmental random vibrations that are typically below 50 kHz. With simulated Q_{anchor} above 19 million, the overall Q of the gyroscope is limited by TED. By eliminating thin beams, this limit is raised to ~180k, which leads to a higher mode-matched SF (Fig. 5.1).

H. Wen, *Toward Inertial-Navigation-on-Chip*, Springer Theses,
https://doi.org/10.1007/978-3-030-25470-4_5

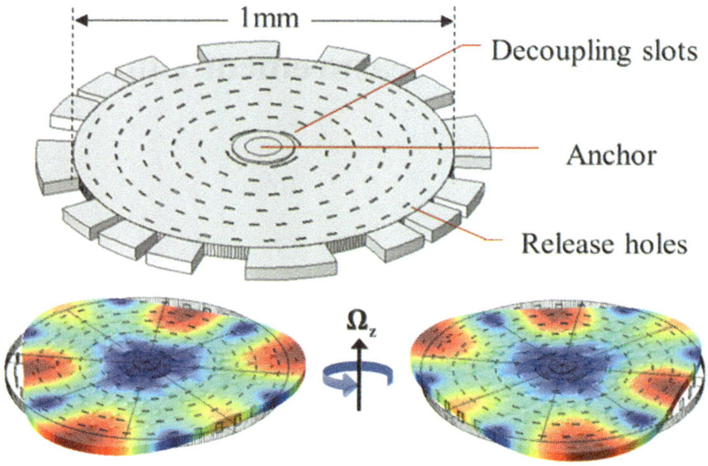

Fig. 5.1 Schematics of the quasi-solid disk structure and the $n = 3$ degenerate BAW gyroscopic modes [1]

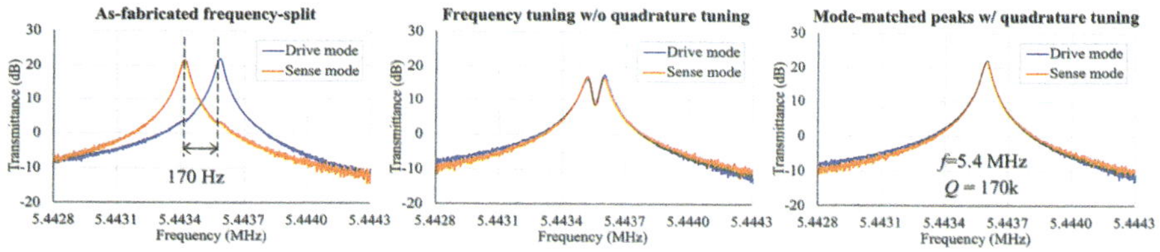

Fig. 5.2 Measured mode-matching behavior of a typical quasi-solid disk yaw gyroscope [1]

The finished design of the quasi-solid disk yaw gyroscope has an operational frequency of around 5.4 MHz. Mode-matching of fabricated parts can be achieved using both frequency tuning and quadrature tuning electrodes with 270 nm gaps. The mode-matches peaks show Q of 170k, which agrees with the simulated TED limit (Fig. 5.2).

5.2 Timing and Inertial Measurement Unit

A single-chip TIMU is designed with robust 3-axis high-frequency resonant gyroscopes, 3-axis high-bandwidth accelerometers, and a high-fQ BAW timing resonator integrated on a 4.5 mm × 5.5 mm platform. The TIMU is fabricated using the HARPSS+ process with wafer-level packaging. The resonant devices (i.e., gyroscopes and resonator) and stationary accelerometers are enclosed in separate cavities

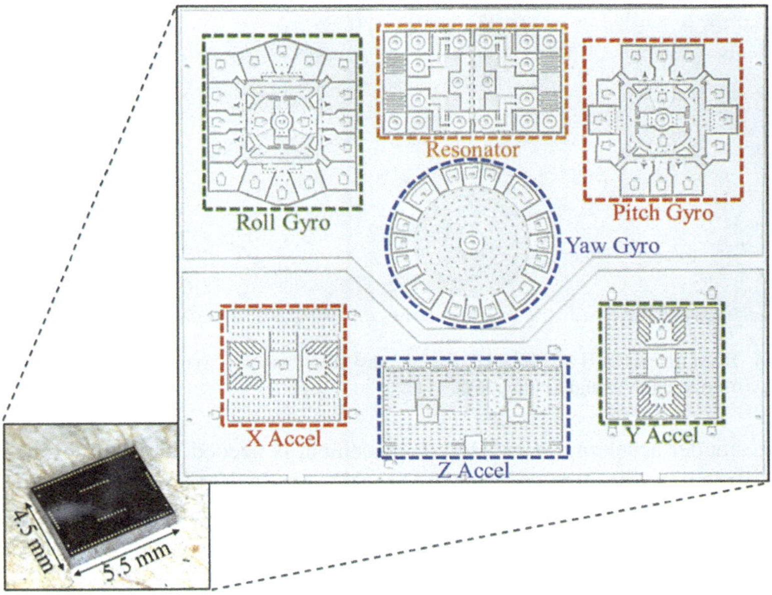

Fig. 5.3 Image of wafer-level-packaged TIMU and SEM images of the uncapped TIMU die [1]

by the bonding, and the accelerometer cavity is brought to atmosphere pressure during dicing through vent holes located at the edge of the cavity to ensure over-damped stable quasi-static response (Fig. 5.3).

5.2.1 3-Axis Gyroscopes

The 3-axis gyroscopes are implemented using one quasi-solid disk BAW gyroscopes and two framed-annulus gyroscopes orientated in orthogonal directions, making it the first single-chip IMU with robust high-frequency resonant gyroscopes for all three sensing axes. Two slightly varied designs are used for pitch and roll sensing with 40 kHz separation between their operational frequencies to reduce cross talk. The handle layer and the substrate are connected together using through BOX layer poly plugs to ensure proper grounding and AC signal isolation (Fig. 5.4).

5.2.2 3-Axis Accelerometers

Nano-gap electrodes enabled by HARPSS+ process are used in the accelerometers. The nano-gaps can significantly improve the electrode transduction and capacitive sensitivity of quasi-static sensors. Therefore, to generate the same level of sensing

Fig. 5.4 SEM images of HARPSS+ fabricated quasi-solid disk yaw gyroscope and framed-annulus pitch gyroscope on the single-chip TIMU [1]

signal, smaller acceleration-induced displacement is needed. This allows the use of stiffer structure with high resonant frequencies above 10 kHz, therefore enables a large bandwidth. The x-axis and y-axis accelerometers have sloped sensing electrodes, which provide larger than gap-size travel range and enables the use of regular (i.e., non-sloped) nano-gaps as shock stops [3, 4]. The z-axis accelerometer uses a seesaw torsional design, which increases the yield and immunity against stiction [5]. In addition, the accelerometer has slanted edge which causes the center of mass to shift lower. This shift counteracts the effect of attaching polysilicon electrodes on top of the device for differential sensing, aligning the center of mass to the neutral axis of the torsional beams. Therefore, no torque is introduced at the beams with in-plane motions, reducing the cross-axis sensitivity.

5.2.3 *Timing Resonator*

In navigational applications, stable onboard clocks are important for robust IMU operation, where the quality of the clock can be crucial for the gyroscope performance [6]. Conventional solutions are based on individually packaged and assembled timing units and IMUs, which cause undesired size, cost, and reliability penalties. Therefore, having an on-chip robust high-frequency clock is essential for self-sustained high-performance delivery of an IMU. A cross-sectional Lamé mode resonator is integrated on the TIMU as a potential timing reference [7]. The design uses localized cross-sectional mode with wide tethers as low-loss anchors to eliminate any vibration sensitivity, which is in line with the high robustness of the high-frequency gyroscopes. Integrated heaters are built around the resonator to provide the potential of closed-loop ovenization of the TIMU for high-temperature stability [8] (Fig. 5.5).

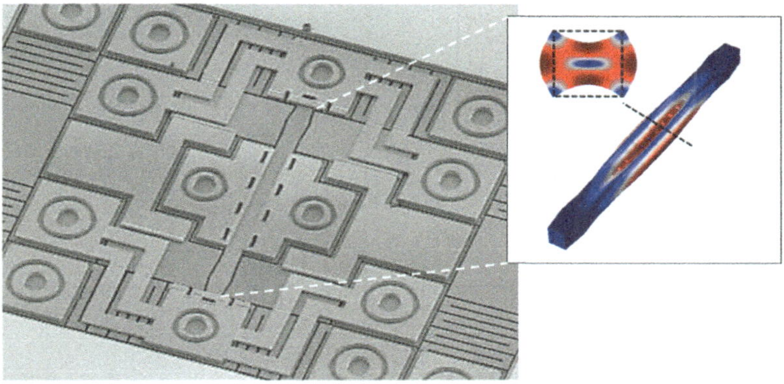

Fig. 5.5 SEM image of the HARPSS+ fabricated cross-sectional Lamé mode resonator with integrated serpentine heaters on the single-chip TIMU. Inset show the mode-shape of the localized cross-sectional Lamé mode [1]

Fig. 5.6 Wire-bonded TIMU die on the evaluation board

5.2.4 *Experimental Characterization*

Fabricated TIMUs are wire-bonded to an evaluation board and characterized with discrete electronics and off-the-shelf commercial ICs (Fig. 5.6). The TIMU dies have a high yield with all devices functional. Figure 5.7 shows the typical response of each devices on a same TIMU die.

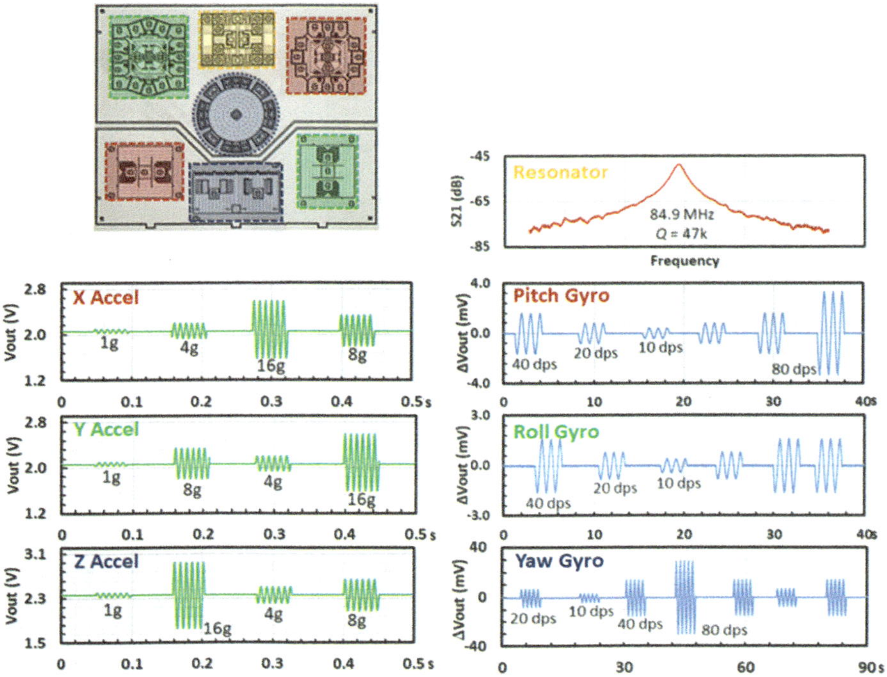

Fig. 5.7 Timing resonator resonant peak and typical time domain response of the IMU subjects to different sinusoidal motions [1]

5.2.4.1 Gyroscope Characterization

Sufficient tuning is achieved to enable mode-matched operation with both yaw and pitch/roll gyroscopes. Yaw gyroscope is mode-matched at 5.43 MHz, showing TED limited Q of 170k for both degenerate modes. Pitch and roll gyros are mode-matched at 660 kHz and 700 kHz, respectively, showing combined TED and SFD limited Q of ~13k for drive mode and ~4.5k for sense mode. Mode-matched gyroscopes are interfaced with simple phase-locked loop and demodulation circuit using a HF2LI lock-in amplifier. Results show large mode-matched SF of 2.7 nA/(°/s) and 0.15 nA/(°/s) with drive amplitude of ~13 nm and ~80 nm for yaw gyroscope and pitch/roll gyroscopes, respectively. The ADEV measurements of the gyroscopes show ARW of 0.23 °/√h and 0.55 °/√h and bias BI of 8.7 °/h and 17.9 °/h for yaw and pitch/roll gyros, respectively. The ARW of yaw gyroscope is limited by the bias instability floor. Once the bias stability is improved, larger drive amplitude can be used along with lower noise electronics to bring down the ARW to a theoretical mechanical limit of 0.04 °/√h. The ARW of pitch/roll gyroscope is limited by the quality-factor due to large SFD. Lowering the package pressure with improved wafer capping process to eliminate the SFD limit can bring the pitch/roll gyroscopes to a theoretical ARW of 0.1 °/√h.

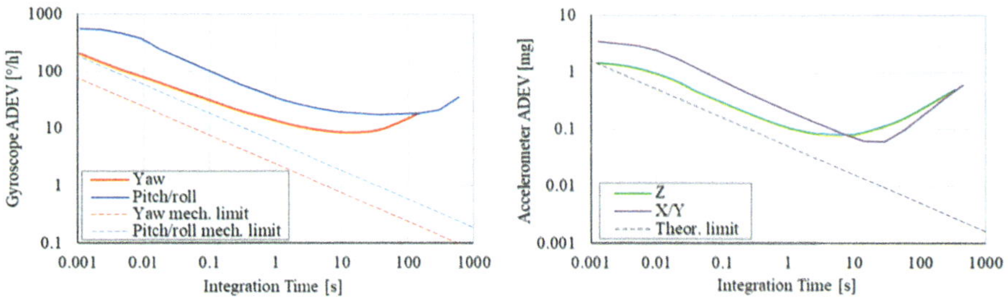

Fig. 5.8 ADEV measurements for gyroscopes and accelerometers on the TIMU [1]

5.2.4.2 Accelerometer Characterization

The translational mode frequencies of the nano-gap accelerometers are measured to be above 10 kHz, confirming a large bandwidth. Testing the accelerometers with off-the-shelf MS3110 ICs gives SF of 28.19 mV/g for x/y accelerometers and 70.29 mV/g for z accelerometer. Cross-axis sensitivity is characterized by manually aligning the TIMU die in different orthogonal directions. Results show less than 5% cross-axis sensitivity, which is limited by the alignment accuracy of the testing setup. ADEV measurements show VRW of 218 μg/√Hz and 85 μg/√Hz and BI of 60 μg and 82 μg for *x*/*y* and *z* accelerometers, respectively (Fig. 5.8).

5.2.4.3 Resonator Characterization

The resonator is found to have a high *Q* of 47k at a high frequency of 84.9 MHz with a linear temperature coefficient of frequency (TCF). This high-*Q* can enable high-stability clock along with ovenization control using the integrated heaters. The possibility of chip-level ovenization also promises lower temperature caused drift in gyroscopes and allows better long-term stability of the TIMU.

5.2.4.4 Summary

The characterized TIMU performance is summarized in Tables 5.1, 5.2, and 5.3. The TIMU demonstrates self-sustained high performance from all sensing axes within a compact form that is suitable for portable devices. It shows, for the first time, the possibility of using 3-axis robust high-frequency resonant gyroscopes to achieve tactical-grade performance, suitable for short-term navigational applications. The performance limits of the current designs will be discussed in the next chapter, which will show that ADEV performance improvement by more than 5~10× is possible through optimizing both the device designs and interface electronics. In addition, the use of high-frequency resonant gyroscopes alone with chip-level ovenization capability brings a lot of potentials and possibilities to the TIMU such as enabling temperature insensitive and low power IMU, which will be presented in Chap. 7.

Table 5.1 Performance summary of the 3-axis gyroscopes on the single-chip TIMU [1]

Parameter	Pitch/roll	Yaw
Operational frequency (MHz)	0.7/0.66	5.4
Q-factors ($\times 10^3$)	13 (drive) 4.5 (sense)	170
Scale-factor (nA/°/s)	0.15	2.7
ARW (°/√h)	0.55	0.23
Bias instability (°/h)	17.9	8.7
Mech. limit (°/√h)	0.1	0.04

Table 5.2 Performance summary of the 3-axis accelerometer on the single-chip TIMU [1]

Parameter	*X*/*Y*	*Z*
Translational *f* (kHz)	14.8	12.5
Scale-factor (mV/g)	28.19	70.29
Cross-axis SF (%)	<4.2	<2.8
VRW (μg/√Hz)	218	85
Bias instability (μg)	60	82
Theo. limit (μg/√Hz)	50	50

Table 5.3 Performance summary of the timing resonator on the single-chip TIMU [1]

Parameter	Value
Frequency (MHz)	84.9
Q-factor ($\times 10^3$)	47
TCF (ppm/°C)	−27

References

1. H. Wen, A. Daruwalla, Y. Jeong, P. Gupta, J. Choi, C.S. Liu, F. Ayazi, A high-performance single-chip timing and inertial measurement unit with robust mode-matched gyroscopes, in *Proceedings of the 31st IEEE International Micro Electro Mechanical Systems Conference (MEMS 2018)*, (IEEE, Belfast, UK, 2018), pp. 105–108. © 2018 IEEE. Figures and tables reprinted with permission
2. R. Mirjalili et al., Substrate-decoupled silicon disk resonators having degenerate gyroscopic modes with Q in excess of 1-million, in *2015 Transducers – 2015 18th International Conference on Solid-State Sensors, Actuators and Microsystems (TRANSDUCERS)* (IEEE, Anchorage, USA, 2015), pp. 15–18
3. Y. Jeong et al., Shock-protection of nano-gap capacitive MEMS accelerometers using sloped electrode design, in *2017 IEEE 30th International Conference on Micro Electro Mechanical Systems (MEMS)* (IEEE, Las Vegas, USA, 2017), pp. 1150–1153
4. Y. Jeong, D.E. Serrano, F. Ayazi, Low-pressure wafer-level-packaged capacitive accelerometers with high dynamic range and wide bandwidth using nano-gap sloped electrode design. J. Microelectromech. Syst. **26**(6), 1335–1344 (2017)
5. Y. Jeong et al., An out-of-plane "hinge-shaped" nano-gap accelerometer with high sensitivity and wide bandwidth, in *2017 19th International Conference on Solid-State Sensors, Actuators and Microsystems (TRANSDUCERS)* (IEEE, Kaohsiung, Taiwan, 2017), pp. 2131–2134
6. G.K. Balachandran et al., A 3-axis gyroscope for electronic stability control with continuous self-test. IEEE J. Solid State Circuits **51**(1), 177–186 (2016)

7. R. Tabrizian, A. Daruwalla, F. Ayazi, High-Q energy trapping of temperature-stable shear waves with Lamé cross-sectional polarization in a single crystal silicon waveguide. Appl. Phys. Lett. **108**(11), 113503 (2016)
8. C.S. Liu, R. Tabrizian, F. Ayazi, Temperature compensated MEMS oscillator using structural resistance based temperature sensing, in *2015 IEEE Sensors* (IEEE, Busan, South Korea, 2015), pp. 1–4

Chapter 6
Bias Stability Limit in Resonant Gyroscopes

The prototype 3-axis resonant MEMS gyroscopes developed in this work demonstrate high performances that are comparable with state-of-the-art commercial parts while providing much better environmental robustness and performance consistency with a small form factor. While promising performance has been achieved for short-term or GPS-aided inertial navigation applications, studying the source of errors and limiting factors in current designs can provide insight into how to further improve the performance and turn the prototypes into products that are feasible for applications with higher performance requirements.

6.1 Electronics and Instrument Limit

The high-Q resonant quasi-solid disk gyroscope demonstrated an ARW among the best ever delivered by MEMS gyroscopes, yet the design still has potential for ~5× ARW improvement to provide near-inertial-grade resolution for very-high-requirement applications such as high-precision scientific measurements. Discussions in Chap. 2 show the white noise-caused ARW of a gyroscope reduces linearly with the drive amplitude. However, bias instability due to flicker noise up-conversion and deterministic bias drift do not scale linearly with drive amplitude. Once the white noise equivalent rotation rate output is reduced to a level comparable with the flicker noise or drift errors, the ADEV value at 1 s will start to be limited by the bias stability, and further increasing the drive amplitude will not improve the actual resolution.

As shown in Fig. 6.1, when the drive amplitude is small, the ADEV follows the $-1/2$ slope line till beyond 1 s, and the ARW reduces as the drive amplitude increases. However, when the drive amplitude is above 10 nm, the ADEV plot starts to deviate from the $-1/2$ slope before 1 s, and the 1 s ADEV value reaches a lower limit with drive amplitude larger than ~13 nm. Further increasing drive amplitude not only doesn't improve the ARW but also can lead to degradation in BI as the

H. Wen, *Toward Inertial-Navigation-on-Chip*, Springer Theses,
https://doi.org/10.1007/978-3-030-25470-4_6

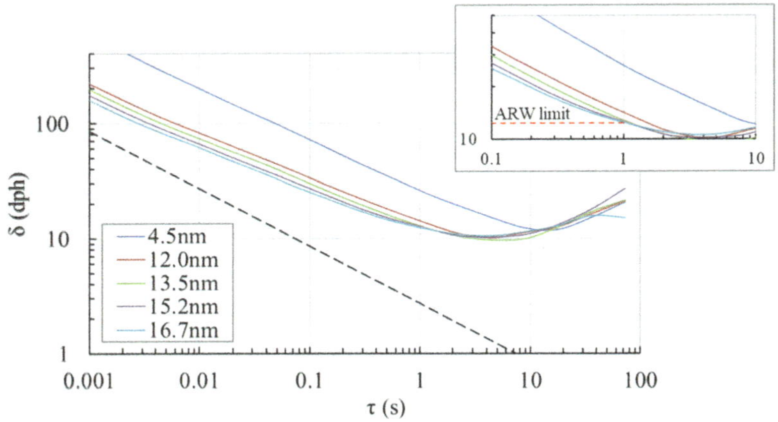

Fig. 6.1 The quasi-solid disk gyroscope ARW scaling with drive amplitude and 1 s resolution limitation due to bias instability. The black dashed line indicates the theoretical mechanical ARW limit with 10% gap-size drive amplitude

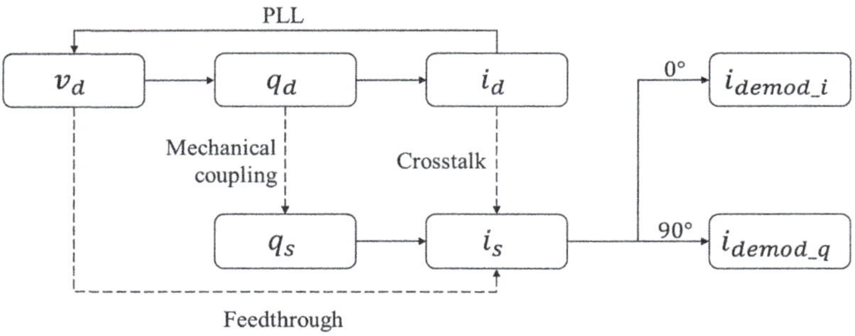

Fig. 6.2 Signal propagation in MEMS gyroscope

drive-loop electronics getting closer to saturation and becoming more nonlinear. To find ways for further resolution improvement, the limiting factors for bias stability needs to be investigated.

The bias in gyroscope comes from the cross-coupling between drive channel and sense channel. As shown in Fig. 6.2, the actuation voltage excites the drive-mode motion which is read out as electrical signal after passing the drive channel front-end electronics. The drive-mode motion is coupled to sense-mode motion through stiffness and damping coupling and read out as electrical signal after passing sense channel front-end electronics. In addition, drive channel electrical signal can be coupled to the sense channel output electrical signal through feedthrough and electronics cross talk. The total sense channel output is then demodulated into the in-phase bias and quadrature error. Instability of the actuation voltage, readout electronics, and demodulation phase can all cause bias instability.

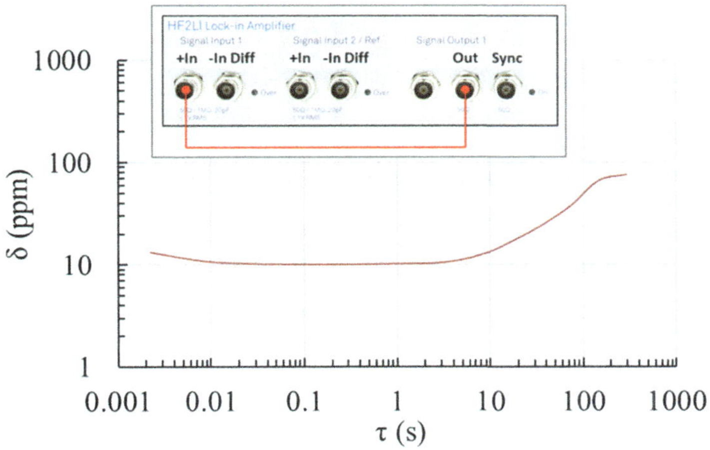

Fig. 6.3 Actuation voltage ADEV measurement without AGC

A major source of flicker noise and environmental drift could be errors from the testing instrument including the actuation and readout interface electronics. To evaluate the contribution of electronics and instrument errors, the measurement setting is examined without connecting the gyroscope in the system. First, the drive signal quality is characterized by directly measuring the actuation voltage output of the lock-in amplifier at a fixed frequency close to the gyroscope resonant frequency. Figure 6.3 shows that a flicker floor corresponding to ~10 ppm relative amplitude instability is found for the actuation voltage signal, which is highly undesirable as errors in the actuation voltage will propagate to every stage in the system and eventually show up in the sense output bias signal.

To reduce errors in the drive voltage of a gyroscope, an AGC is usually implemented using the gyroscope drive-mode output as a feedback signal to stabilize the displacement amplitude. Beside drive voltage, noise and errors exist in the front-end electronics and readout instrument. This is captured by measuring the circuit output while replacing the gyroscope on the evaluation board with a low-noise low-TCR (temperature coefficient of resistance) thin-film resistor (RNCF0603TKY10K0) with value similar to the gyroscope motional impedance. The sync signal (proportional to the output voltage) of the drive voltage is fed into a second channel for AGC implementation to ensure the electronics are driven by a clean and stable voltage. Figure 6.4 shows a typical result of the experiment. For a clean input signal, a relative readout gain instability above ~6 ppm is found due to errors in the readout circuit and instrument.

Both errors in the actuation voltage and interface electronics will contribute to the sense output error through the in-phase bias of the gyroscope. If no AGC is used, the total rotation rate output bias error cause by the uncorrelated noise in the drive voltage and readout electronics is:

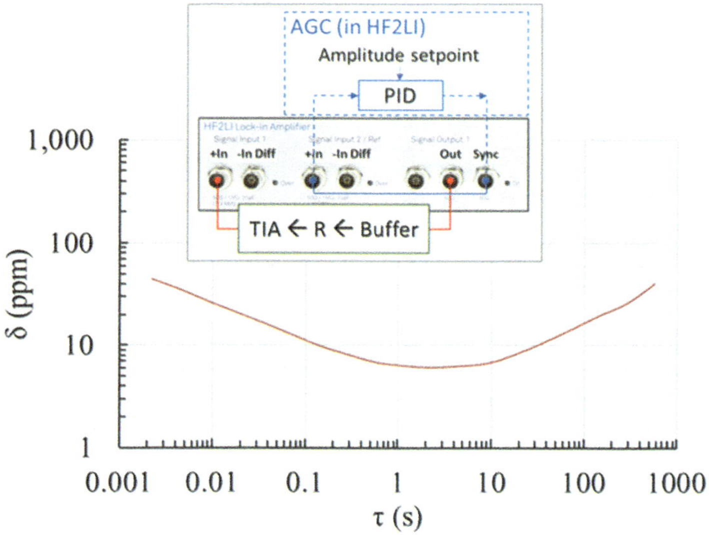

Fig. 6.4 Readout circuit and instrument ADEV measurement

$$\delta B_{\Omega_{\text{in-phase}}} = B_{\Omega_{\text{in-phase}}} \cdot \sqrt{\left(\frac{\delta v_d}{v_d}\right)^2 + \left(\frac{\delta A_{\text{elec}}}{A_{\text{elec}}}\right)^2}, \tag{6.1}$$

where the $B_{\Omega_{\text{in-phase}}}$ is the in-phase bias equivalent rate output level in unit of (°/h) and $\delta v_d/v_d$ and $\delta A_{\text{elec}}/A_{\text{elec}}$ are the relative errors in the actuation voltage and error in readout gain as described before.

When characterizing a gyroscope, both channels of the lock-in amplifier will be used to readout the signal from drive mode and sense mode of the gyroscope. Therefore, when AGC is to be used, instead of using a dedicated channel for AGC implementation, the drive-oscillation-loop readout will be used as the feedback signal. As a result, errors in the drive-loop front-end electronics will be passed onto the drive voltage. Assuming the drive channel and sense channel has similar uncorrelated electrical noise, the total rotation rate output bias error is:

$$\begin{aligned}\delta B_{\Omega_{\text{in-phase}}} &\approx B_{\Omega_{\text{in-phase}}} \cdot \sqrt{\left(\frac{\delta A_{\text{elec\_drive}}}{A_{\text{elec\_drive}}}\right)^2 + \left(\frac{\delta A_{\text{elec\_sense}}}{A_{\text{elec\_sense}}}\right)^2} \\ &= \sqrt{2} B_{\Omega_{\text{in-phase}}} \frac{\delta A_{\text{elec}}}{A_{\text{elec}}}.\end{aligned} \tag{6.2}$$

Equations (6.1) and (6.2) show the bias instability of the gyroscope due to electronics, and instrument errors are proportional to the in-phase bias level. Large bias level due to damping coupling causes high BI when combined with electronics and instrument errors. In the quasi-solid disk gyroscope, substrate decoupling is realized by shielding slots to balance the acoustic waves at the anchor to visually isolate the resonant modes. A drawback of the design is the decoupling slots are sensitive to fabrication variations such as trench size variation and disk ovalness, which will cause principle damping axis misalignment and result in a large damping coupling. For the device characterized, a bias level of around 10^{6}°/h is measured. Based on (6.2), the ADEV contribution due to the electronics and instrument errors is calculated and compared with the overall gyroscope performance. As shown in Fig. 6.5, the overall long-term bias instability of the quasi-solid disk gyroscope is dominated by the electronics and instrument error. The 1 s limit is slightly higher than the predicted electronics contribution. The possible reason is as follows: For smaller drive amplitude, the 1 s value is a combined result of electronics errors and gyroscope white noise. As the drive amplitude increases, the white noise contribution reduces, but the drive channel electronics becomes more nonlinear as the output swing is getting closer to the saturation point of the electronics, causing increased noise up-conversion and instrument errors. Consequently, $\delta A_{elec\_drive}/A_{elec\_drive}$ becomes larger than $\delta A_{elec\_sense}/A_{elec\_sense}$, and the total contribution becomes larger than the value given by (6.2). This can also be seen from the increased BI between 1 and 10 s for drive amplitude larger than 13.5 nm. Therefore, to achieve further performance improvement, modifications must be made to the substrate-decoupling mechanism to improve the process variation immunity and to achieve lower in-phase bias level. Once the bias level is reduced, larger drive amplitude can be used, and both ARW and BI will be improved.

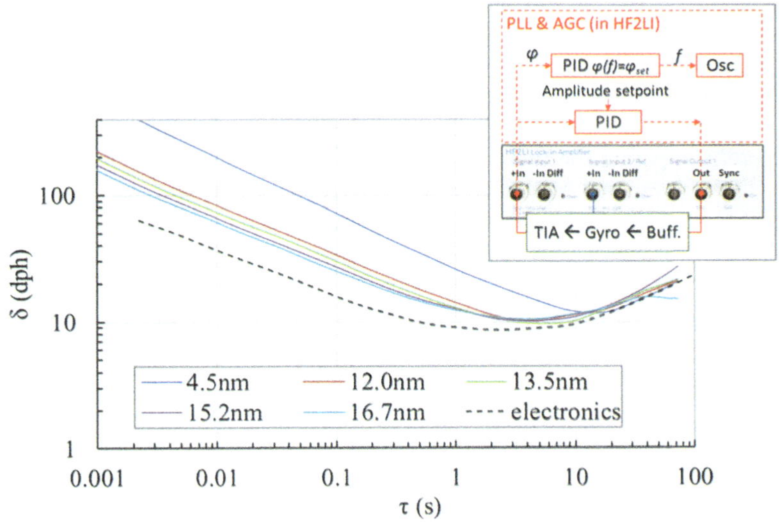

Fig. 6.5 Comparison between the measured gyroscope performance and the estimated electronics and instrument error contribution based on one separate electronics characterization results

6.2 Demodulation Phase Error

To verify the findings about high-bias-level related instrument limit and to identify other possible performance limits, a different BAW yaw gyroscope is also characterized. The gyroscope uses a serpentine thin beam network design to realize substrate decoupling. The gyroscope has the operational modes at ~4.2 MHz with Q of 100k. Although the use of thin beam tethers compromised their translational motion resistance and quality-factor, it allows the resonant structure to be fully isolated from the anchor in a way that is insensitive to lithography errors and other process variations and therefore having lower damping coupling. Measurement shows this design enables a much better drive mode and sense-mode isolation with typical in-phase bias levels below 3×10^5°/h.

Two implementations of the same gyroscope design are characterized. One is implemented on a 6-DOF inertial sensor die with metal traces for electrical access to the electrodes, and the other one is an individual device with metal pads directly on top of the electrodes (Fig. 6.6).

The device on 6-DOF is first characterized. Despite having a low in-phase bias level, a BI of 11.6 °/h is found, and with drive amplitude ~15 nm, the ADEV is deviated from the −1/2 slope, giving an ARW of 0.35 °/√h. This result indicates that errors other than the electronics noise exist which are limiting the performance of the substrate-decoupled gyroscope. Characterization of the same design on an individual die with same actuation level, however, shows a well-behaved ARW and a lower BI. The ADEV of the individual device follows the −1/2 slope till more than 1 s, giving an ARW of 0.29 °/√h and BI of 7 °/h. Higher performance is achieved with larger drive amplitude, and the ADEV starts to deviate from the −1/2 slope at 1 s for drive amplitude above 22 nm (8% gap-size), giving an ARW of 0.19 °/√h and BI of 5.5 °/h. While the two devices present the same Q, SF, and ADEV at low averaging time with the same drive amplitude, a clear difference in their BI can be seen. Therefore, comparing the two devices may reveal fundamental root of the performance limit in resonant MEMS gyroscopes and provide insight into improving the BAW yaw gyroscope as well as high-frequency pitch and roll gyroscopes (Fig. 6.7).

A major difference between the two devices is their electrical connections. Although the electrode configurations are the same, the device on the 6-DOF die is connected through long metal traces with drive and sense signals routed next to

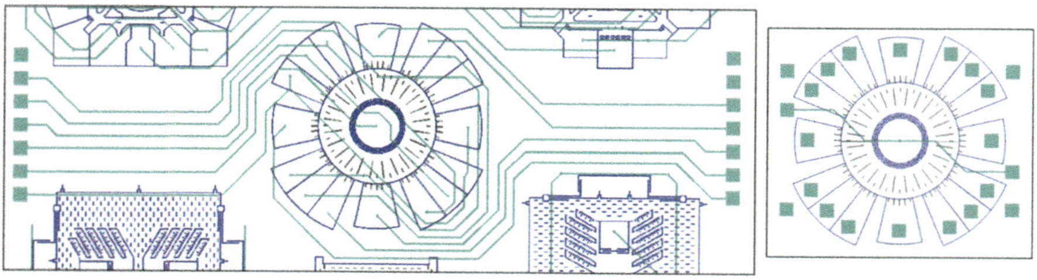

Fig. 6.6 Layout images of the substrate-decoupled yaw gyroscopes. (Left) device with metal traces. (Right) device with metal directly on top of electrodes

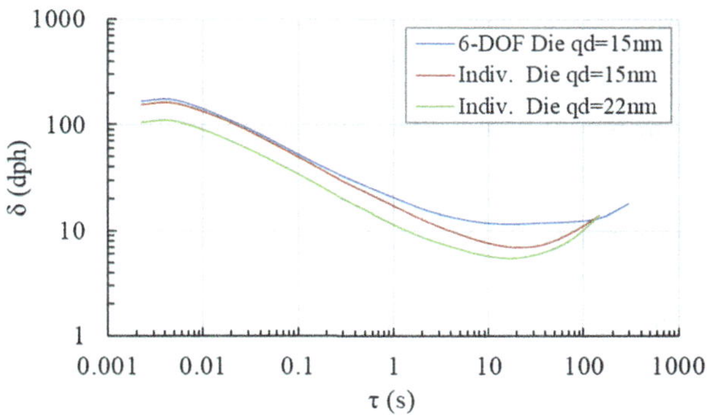

Fig. 6.7 ADEV measurements of the substrate-decoupled BAW yaw gyroscopes on a 6-DOF die and on an individual die

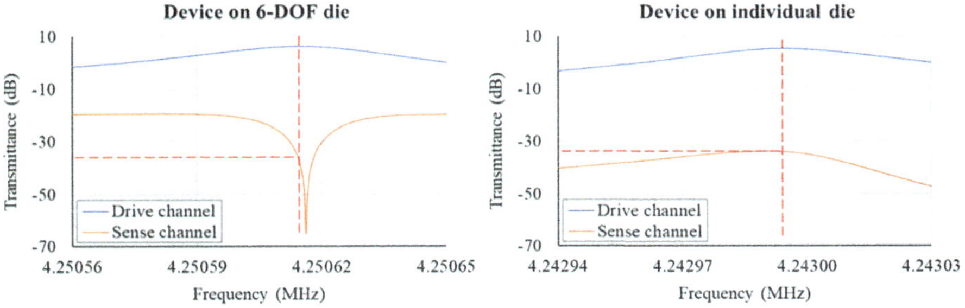

Fig. 6.8 Frequency response of the BAW yaw gyroscope after adjusting tuning voltages to match the frequency and null the quadrature channel. The remaining sense channel output level at the resonant frequency corresponds to the in-phase bias level

each other. A possible effect of this is a larger feedthrough and electrical coupling between the drive and sense channels. As pointed out in Chap. 2, the feedthrough signal is in quadrature phase for a resonant gyroscope, and an excessive amount of feedthrough can lead to inaccurate quadrature cancellation. This effect is observed in the frequency response of the two devices. As shown in Fig. 6.8, a feedthrough floor of about −20 dB is found in the sense channel for the device with the metal traces, while the device with direct pads has a feedthrough level less than −40 dB. Ideally, during mode-matching, the demodulated quadrature output is tuned to close to zero to eliminate stiffness cross-coupling, leaving only the in-phase bias component caused by damping coupling. However, according to (2.64), making the demodulated quadrature output signal gives:

$$\left|i_{\text{demod}_q}\right| \approx \left|i_{\text{mech}_q} + i_{\text{elec}_q}\right| = 0 \rightarrow \left|i_{\text{mech}_q}\right| = \left|i_{\text{elec}_q}\right|. \tag{6.3}$$

When close to the sense-mode resonant frequency, this results in a residual mechanical quadrature corresponding to an output level similar to the feedthrough level. Therefore, even though the same extent of quadrature nulling and similar in-phase bias levels are obtained in time domain response for both devices, a much larger unobserved mechanical quadrature exists in the device with the metal traces than the device with direct metal pads. According to (2.63), with the same amount of demodulation phase instability, a much larger in-phase bias variation will be caused in the device with higher feedthrough, resulting in the larger BI as observed in the ADEV measurements.

For yaw gyroscopes using degenerate modes, the output bias given in (2.32) can be written as:

$$|B_i| = \frac{|\omega_0 q_1 d_{12}|}{\sqrt{\left(k_{22} - \omega_0^2 m\right)^2 + (\omega_0 d_{22})^2}} \text{ and } \left|B_q\right| = \frac{|q_1 k_{12}|}{\sqrt{\left(k_{22} - \omega_0^2 m\right)^2 + (\omega_0 d_{22})^2}} \tag{6.4}$$

$$\varphi_i = \angle \frac{B_i}{q_1} = \tan^{-1}\left(\frac{k_{22} - \omega_0^2 m}{\omega_0 d_{22}}\right). \tag{6.5}$$

For high-Q gyroscope with close to mode-matched operation, substitute in (2.21) and (2.23):

$$\left|\frac{k_{22} - \omega_0^2 m}{\omega_0 d_{22}}\right| \ll 1 \rightarrow \varphi_i \approx \frac{k_{22} - \omega_0^2 m}{\omega_0 d_{22}} = -\frac{1}{\omega_0} \frac{(k_1 - k_2)(1 + \cos 2\theta_\omega)}{d_1 + d_2 - (d_1 - d_2)\cos 2\theta_\tau}. \tag{6.6}$$

From (6.6) we can see that the correct demodulation phase depends on both the principle stiffness (or frequency) and the principle damping constants. Instability and flicker noise behavior in the resonant frequency of MEMS resonators have been widely reported, and most dissipation mechanisms in MEMS resonators are known to have environmental dependence and could drift over time with environmental variations.

For a qualitative study, assuming the actuation frequency is locked to the observed drive-mode resonant frequency (i.e., $\omega_0 = (k_1/m)^{1/2}$), when there are fluctuations in the principle resonant frequencies or principle dissipations, (6.6) gives:

$$\begin{aligned}
\delta\varphi_{k_1} &= \frac{\partial \varphi_i}{\partial k_1}\delta k_1 = \varphi_i \frac{k_1 + k_2}{2k_1}\frac{1}{k_1 - k_2}\delta k_1 \\
\delta\varphi_{k_2} &= \frac{\partial \varphi_i}{\partial k_2}\delta k_2 = \varphi_i \frac{-1}{k_1 - k_2}\delta k_2 \\
\delta\varphi_{d_1} &= \frac{\partial \varphi_i}{\partial d_1}\delta d_1 = \varphi_i \left[-\frac{1 - \cos 2\theta_\tau}{d_1 + d_2 - (d_1 - d_2)\cos 2\theta_\tau}\delta d_1\right] \\
\delta\varphi_{d_2} &= \frac{\partial \varphi_i}{\partial d_2}\delta d_2 = \varphi_i \left[-\frac{1 + \cos 2\theta_\tau}{d_1 + d_2 - (d_1 - d_2)\cos 2\theta_\tau}\delta d_2\right].
\end{aligned} \tag{6.7}$$

Considering a close to mode-matched operation with small damping mismatch, (6.7) gives:

$$
\begin{aligned}
\delta\varphi_{k_1} &\approx \varphi_i \frac{1}{k_1 - k_2}\delta k_1 = \varphi_i \frac{1}{2\Delta k}\delta k_1 \\
\delta\varphi_{k_2} &\approx -\varphi_i \frac{1}{k_1 - k_2}\delta k_2 = -\varphi_i \frac{1}{2\Delta k}\delta k_2 \\
\delta\varphi_{d_1} &\approx -\varphi_i \frac{1 - \cos 2\theta_\tau}{d_1 + d_2}\delta d_1 = -\varphi_i \frac{1 - \cos 2\theta_\tau}{2d}\delta d_1 \\
\delta\varphi_{d_2} &\approx -\varphi_i \frac{1 + \cos 2\theta_\tau}{d_1 + d_2}\delta d_2 = -\varphi_i \frac{1 + \cos 2\theta_\tau}{2d}\delta d_2 .
\end{aligned}
\tag{6.8}
$$

where Δk and d are defined to be $(k_1 - k_2)/2$ and $(d_1 + d_2)/2$, respectively. When considering uncorrelated noise and assuming the two modes have similar noise density:

$$
\begin{aligned}
|\delta\varphi| &= \sqrt{(\delta\varphi_{k_1})^2 + (\delta\varphi_{k_2})^2 + (\delta\varphi_{d_1})^2 + (\delta\varphi_{d_2})^2} \\
&= |\varphi_i| \sqrt{\left[\frac{(\delta k_1)^2 + (\delta k_2)^2}{(2\Delta k)^2}\right] + \left[\frac{(1 - \cos 2\theta_\tau)^2(\delta d_1)^2 + (1 + \cos 2\theta_\tau)^2(\delta d_2)^2}{(2d)^2}\right]} \\
&= \frac{|\varphi_i|}{\sqrt{2}} \sqrt{\left(\frac{\delta k}{\Delta k}\right)^2 + \left[(1 + \cos^2 2\theta_\tau)\frac{\delta d}{d}\right]^2} .
\end{aligned}
\tag{6.9}
$$

For resonant operation, we have $\Delta k \to 0$. Therefore, the first term will be much larger than the second term, and the demodulation phase error will be dominated by the frequency noise of the two principle resonant modes. However, when considering deterministic errors (e.g., due to ambient temperature variations):

$$
\begin{aligned}
\delta\varphi &= \delta\varphi_{k_1} + \delta\varphi_{k_2} + \delta\varphi_{d_1} + \delta\varphi_{d_2} \\
&= \varphi_i \left[\frac{\delta k_1 - \delta k_2}{2\Delta k} - \frac{(1 - \cos 2\theta_\tau)\delta d_1 + (1 + \cos 2\theta_\tau)\delta d_2}{2d}\right] \\
&\approx \varphi_i \left[\frac{\delta(\Delta k)}{\Delta k} - \frac{\delta d}{d}\right] .
\end{aligned}
\tag{6.10}
$$

The first term represents the relative frequency change between the drive and sense modes, and the second term represents the dissipation variation in the gyroscope. Which term contributes more depends on the design and root of variation. For example, for ambient temperature variations, if degenerate modes or modes with similar TCF are used, the contribution from the first term is usually small compared to the second term. But for substrate stress variations, either term can dominate depending on the designs.

The frequency and dissipation variations also cause changes in the drive-mode phase response. In practice, the phase-locked loop will react to these variations and ensure a fixed drive-loop output phase with an actuation frequency slightly off the observed drive-mode frequency, which also reduces the sense output phase error. A quantitative evaluation of the total phase error and the effect on bias stability can be done using numerical models based on (2.19) and related equations defined in Chap. 2. In addition to phase shifts in the mechanical response, phase errors from the TIAs and other electronics will cause additional demodulation errors. At an operational frequency of ~4 MHz, the TIAs used are measured to have a phase instability of a few μrads. The total phase error due to different error sources is then given by:

$$|\delta\varphi_{\text{total}}| = \sqrt{(\delta\varphi_{\text{mech}})^2 + (\delta\varphi_{\text{elec}})^2} \approx \sqrt{\left(\frac{\varphi_i}{\sqrt{2}\Delta k}\delta k - \delta\varphi_{\text{PLL}}\right)^2 + (\delta\varphi_{\text{elec}})^2} \quad (6.11)$$

where $\delta\varphi_{\text{PLL}}$ is the sense-mode phase shift compensation due to drive-loop PLL control and $\delta\varphi_{\text{elec}}$ is the phase error in the interface electronics.

Beside demodulation phase error, variations in the resonator parameters can also affect the cross-coupling amplitude. Again, assuming close to mode-matched condition:

$$|B_i| \approx \left|\frac{q_1(d_1 - d_2)\sin 2\theta_\tau}{d_1 + d_2 - (d_1 - d_2)\cos 2\theta_\tau}\right| \quad (6.12)$$

$$|B_q| \approx \left|\sqrt{\frac{m}{k_1}}\frac{q_1(k_1 - k_2)\sin 2\theta_\omega}{d_1 + d_2 - (d_1 - d_2)\cos 2\theta_\tau}\right|. \quad (6.13)$$

Considering uncorrelated noise, taking the derivatives of (6.12) and (6.13) and assuming small damping mismatch give:

$$|\delta B_{i_{\text{amp}}}| \approx |B_i|\sqrt{\frac{(\delta d_1)^2 + (\delta d_2)^2}{(d_1 - d_2)^2}} \approx \frac{1}{\sqrt{2}}\left|B_i\frac{\delta d}{\Delta d}\right| \quad (6.14)$$

$$|\delta B_{q_{\text{amp}}}| \approx \frac{|B_q|}{\sqrt{2}}\sqrt{\left(\frac{\delta k}{\Delta k}\right)^2 + \left[(1 + \cos^2 2\theta_\tau)\frac{\delta d}{d}\right]^2} \approx \frac{1}{\sqrt{2}}\left|B_q\frac{\delta k}{\Delta k}\right|, \quad (6.15)$$

where $\Delta d = (d_1 - d_2)/2$. When the two modes are tuned to the minimum frequency split (veering point), we have $\theta_\omega = 45°$ as given in (2.53). For small damping mismatch:

$$
\begin{aligned}
|B_i| &\approx \left|\frac{q_1 \Delta d \sin 2\theta_\tau}{d}\right| \\
|B_q| &\approx \left|\frac{q_1 \Delta k \sin 2\theta_\omega}{\omega_0 d}\right| = \left|\frac{q_1 \Delta k}{\omega_0 d}\right| \\
|\varphi_i| &\approx \left|\frac{\Delta k (1 + \cos 2\theta_\omega)}{\omega_0 d}\right| = \left|\frac{\Delta k}{\omega_0 d}\right|.
\end{aligned} \tag{6.16}
$$

Substituting (6.16) into (6.9), (6.14), and (6.15) gives:

$$
\left|\delta B_{i_{\text{phase}}}\right| = \left|B_q \delta\varphi\right| \approx \left|B_q\right| \sqrt{\left(\frac{1}{\sqrt{2}\omega_0 d}\delta k - \delta\varphi_{PLL}\right)^2 + \left(\delta\varphi_{\text{elec}}\right)^2} \tag{6.17}
$$

$$
\left|\delta B_{q_{\text{phase}}}\right| = \left|B_i \delta\varphi\right| \approx \left|B_i\right| \sqrt{\left(\frac{1}{\sqrt{2}\omega_0 d}\delta k - \delta\varphi_{\text{PLL}}\right)^2 + \left(\delta\varphi_{\text{elec}}\right)^2} \tag{6.18}
$$

$$
\left|\delta B_{i_{\text{amp}}}\right| \approx \frac{1}{\sqrt{2}} \left|\frac{q_1 \sin 2\theta_\tau}{d} \delta d\right| \tag{6.19}
$$

$$
\left|\delta B_{q_{\text{amp}}}\right| \approx \frac{1}{\sqrt{2}} \left|\frac{q_1}{\omega_0 d} \delta k\right|. \tag{6.20}
$$

Two things are worth noting from (6.17), (6.18), (6.19), and (6.20). First, stiffness (frequency) instability δk causes both in-phase channel and quadrature channel errors, whereas damping instability δd only affects the in-phase channel. Second, the in-phase bias instability is proportional to the quadrature level if caused by δk and is independent of the quadrature level if caused by δd, whereas quadrature channel instability is always independent of the quadrature level. Therefore, comparing the stabilities of the in-phase and quadrature channels and how they react to quadrature level changes can help determine the root of the bias instability limit in the gyroscope.

As shown in Fig. 6.9, the individual die with low feedthrough is tested with different intentionally added quadrature levels by changing the quadrature tuning voltage. The results show a quadrature-level-dependent in-phase bias instability and a quadrature-level-independent quadrature bias instability, matching the behavior predicted by (6.17) and (6.20). For a quadrature level of −22 dB, which is similar to the feedthrough level in the device with metal traces, a similar BI limit of 11.3 °/h is found, verifying the demodulation phase error combined with high quadrature residual due to large feedthrough was the limiting factor in the device with metal traces. A numerical model of the gyroscope bias instability is implemented based on (2.19) with measured output bias levels and electronics phase errors (Appendix B.2). A frequency instability of ~0.02 ppb is found to fit well with the measured gyroscope quadrature channel instability and in-phase channel bias instability with high quadrature. The modeled bias instability levels are plotted with dashed lines in Fig. 6.9.

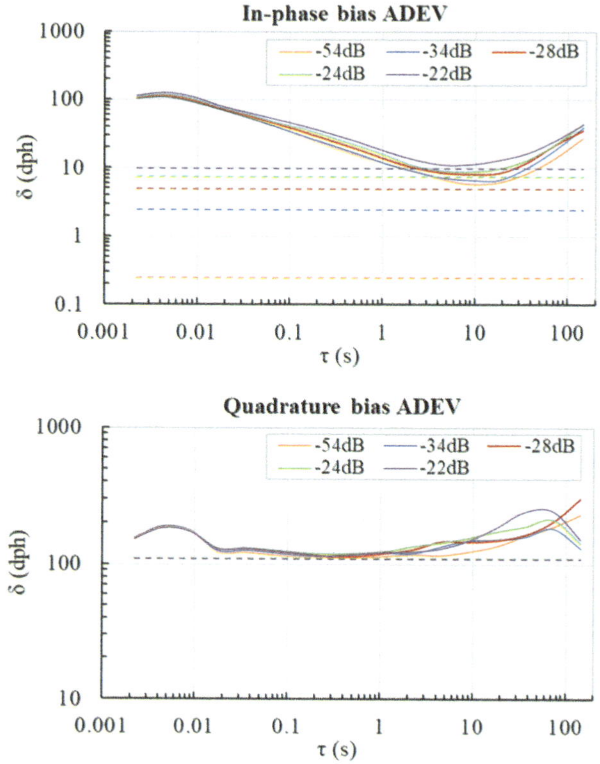

Fig. 6.9 In-phase and quadrature channel ADEV measurements with different quadrature levels. The dashed lines show the BI values predicted by the numerical model

As the quadrature level reduces, both modeled and measured in-phase BI reduces. However, discrepancies are found for low quadrature levels as the measured value reaches a lower limit, while the modeled value scales down linearly with the quadrature level, which indicates errors other than the demodulation phase error have started to limit the gyroscope performance when quadrature is reduced. To identify this limit, the readout electronics and instrument errors are characterized as mentioned in the previous section. As shown in Fig. 6.10, the measured electronics and instrument error limits match well with the overall gyroscope performance both with and without AGC, verifying the gyroscope performance has reached the instrument limit. The numerical model of the gyroscope is modified to include the instrument limit, and the results are listed in Table 6.1. The modified model shows a good agreement with the measured values. Therefore, it can be used in future gyroscope characterizations to quantify the performance limiting factors and provide guidelines for making improvements.

The learnings from the substrate-decoupled BAW yaw gyroscopes can be applied to other high-frequency resonant gyroscopes. Large feedthrough and electronics cross talks are detrimental for gyroscopes with high-operational frequencies as they affect the quadrature cancellation accuracy and result in large demodulation

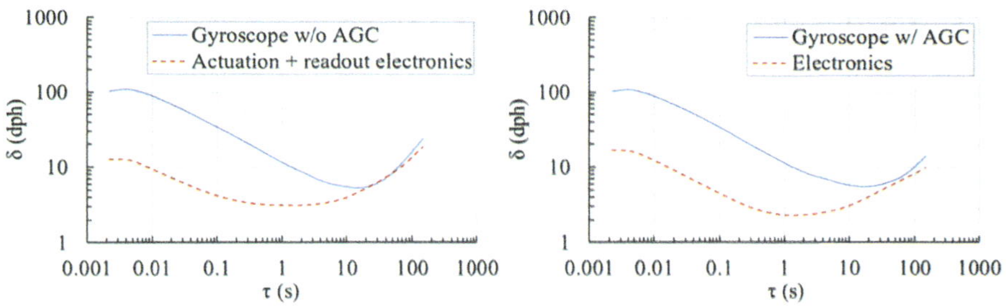

Fig. 6.10 Measured gyroscope ADEV and electronics contributions with and without AGC

Table 6.1 Measured and modeled bias instability of the substrate-decoupled BAW yaw gyroscope

	Quadrature level (dB)	Measurement	Model with freq. instability	Model with freq. and instrument errors
In-phase channel BI (°/h)	−22	11.318	9.969	11.301
	−24	8.996	7.488	9.186
	−28	8.224	4.997	7.300
	−34	6.749	2.500	5.880
	−54	5.849	0.250	5.328
Quadrature channel BI (°/h)	−22	115.84	109.67	116.41
	−24	119.32	109.87	113.70
	−28	113.70	110.01	111.73
	−34	115.32	110.10	110.53
	−54	112.44	110.12	110.13

error-induced bias instability. The effect cannot be easily avoided as the phase relation between the electrical coupling signal and mechanical quadrature can vary due to small frequency mismatch and variations in other operation conditions. Especially for pitch and roll gyroscopes with nondegenerate modes where nonidentical front-end electronics may be used, causing different phase shifts in the drive and sense channels that can make the phase relation between electrical and mechanical coupling signals more complicated. Therefore, cautions need to be taken when designing the gyroscope electrode configurations, connection routings, as well as shielding and isolations on the evaluation board.

When quadrature is properly cancelled, the gyroscopes become limited by the actuation and readout electronics. To minimize the effects of errors in the electronics, a large SF to in-phase bias ratio is desirable, which leads to smaller equivalent rotation rate output errors for the same electronics stability limit. Following (3.10), this can be achieved by using modes with high Coriolis sensitivity, large effective mass, and smaller damping coupling through substrate decoupling and designs with more symmetric damping mechanisms.

6.3 Performance of Pitch or Roll Gyroscopes

The fundamental limits discussed in the previous sections are applicable to all resonant gyroscopes. However, contributions of each factors and design strategies may be different for yaw and pitch or roll gyroscopes due to differences in the mode of choice, operational frequencies, and fabrication feasibilities.

6.3.1 Performance Scaling Comparison

Before this work, high-frequency resonant pitch or roll gyroscope performance has been deficient as compared to yaw gyroscopes. Progresses presented in this dissertation have brought high-frequency resonant pitch or roll gyroscopes to a comparable level as the yaw gyroscopes. While the paths forward are similar for yaw and pitch or roll gyroscopes, different design implementations might need to be taken.

Table 6.2 shows a comparison of the specifications of the quasi-solid disk yaw gyroscope and the framed-annulus pitch or roll gyroscopes. The lower operational frequency of pitch or roll gyroscope should ideally allow lower motional impedance for lower drive power and better ARW with lower electrical noise. However, both parameters are found to be worse in the pitch or roll gyroscope due to lower Q, effective mass, and operation voltages. Due to the lower resonant frequency, the quality-factors of the pitch or roll gyroscopes are limited by SFD. In the future, achieving better cavity vacuum with more advanced WLP (e.g., with getters) will significantly improve the Q-factors, leading to better motional impedance, mechanical SF, and ARW. Furthermore, eliminating SFD will also allow lower damping coupling with no hydrodynamic lift effect and more symmetric dissipations between the drive and sense modes limited by TED, which will lead to lower in-phase bias and better BI.

Besides the lower quality-factors, the SF and noise performance of the pitch or roll gyroscope are also limited by the sense-mode effective mass and applicable polarization voltage. As listed in Table 6.2, the pitch or roll gyroscope occupies a slightly smaller area than the yaw gyroscope, which provides room to increase the effective mass with larger resonator size without causing degradation to the overall form factor when integrated to make IMUs. Due to the thickness limit, the pitch or roll gyroscope presents lower OOP stiffness as compared to the IP stiffness, therefore having a lower pull-in voltage above which the resonator will be stuck to the OOP electrodes. Going to larger device thickness or using stiffer tethers will mitigate the voltage limit, but the associated challenges in fabrication yield and device isolation need to be considered. Another possible solution is using both top and bottom electrodes to generated balanced DC electrostatic force, therefore significantly increasing the pull-in voltage limit. Using both top and bottom electrodes may also allow larger OOP frequency tuning range with smaller electrode pad size and

Table 6.2 Design specification and performance comparisons between the quasi-solid disk yaw gyroscope and framed-annulus pitch or roll gyroscope

Parameter	Quasi-solid disk yaw gyroscope	Framed-annulus pitch/roll gyroscope
Frequency (MHz)	5.4	0.66
Resonator dimensions (mm^2)	1.04 × 1.04	0.92 × 0.89
Effective mass (μg)	26	6.2 (drive) 5.1 (sense)
Coriolis coupling coefficient	0.6	0.8
Q-factors (×10^3)	170	13 (drive) 4.5 (sense)
Pull-in voltage (V)	>50	22
Motional impedance (kΩ)	6	260
Drive amplitude (nm)	13	80
Scale-factor (nA/°/s)	2.7	0.15
ARW by −1/2 slope fit (°/√h)	0.1	0.55
Bias instability (°/h)	8.7	17.9

larger differential output transduction area. However, the implementation of bottom is not straightforward and would require further advancements in the fabrication process.

6.3.2 Nonlinear Tuning Effect

An additional source of error identified in the high-frequency resonant pitch or roll gyroscopes is the nonlinear tuning effect under large drive displacements. The framed-annulus gyroscope uses nano-gap comb-drives with large linear actuation range. However, this causes larger than 1/10 gap-size variations at the slanted quadrature tuning electrodes, and higher-order displacement terms need to be considered in the electrode capacitance:

$$C_s(q_1, q_2) \approx \frac{\varepsilon A_s}{g_0}\left[1 + \frac{\alpha_1 q_1 \sin\theta \pm \alpha_2 q_2 \cos\theta}{g_0} + \left(\frac{\alpha_1 q_1 \sin\theta \pm \alpha_2 q_2 \cos\theta}{g_0}\right)^2 + \left(\frac{\alpha_1 q_1 \sin\theta \pm \alpha_2 q_2 \cos\theta}{g_0}\right)^3 \ldots\right]. \tag{6.21}$$

For the framed-annulus design, the corresponding electrostatic stiffness terms are:

$$\begin{aligned}
k_{e11} &\approx -\frac{\varepsilon A_s V_{DC}^2}{g_0^3}\left(\alpha_1^2 \sin^2\theta + 3\alpha_1^3 \sin^3\theta \frac{q_1}{g_0}\right) = k_{e11_0}\left(1 + 3\alpha_1 \sin\theta \frac{q_1}{g_0}\right) \\
&\approx k_{e11_0}\left(1 + 1.5\frac{q_1}{g_0}\right) \\
k_{e22} &\approx -\frac{\varepsilon A_s V_{DC}^2}{g_0^3}\left(\alpha_2^2 \cos^2\theta + 3\alpha_1\alpha_1^2 \sin\theta \cos^2\theta \frac{q_1}{g_0}\right) = k_{e22_0}\left(1 + 3\alpha_1 \sin\theta \frac{q_1}{g_0}\right) \\
&\approx k_{e22_0}\left(1 + 1.5\frac{q_1}{g_0}\right) \\
k_{e12} &\approx -\frac{\varepsilon A_s V_{DC}^2}{g_0^3}\left(\pm\alpha_1\alpha_2 \sin\theta\cos\theta \pm 3\alpha_1^2\alpha_2 \sin^2\theta \cos\theta \frac{q_1}{g_0}\right) \\
&= k_{e12_0}\left(1 + 3\alpha_1 \sin\theta \frac{q_1}{g_0}\right) \approx k_{e12_0}\left(1 + 1.5\frac{q_1}{g_0}\right).
\end{aligned} \tag{6.22}$$

Equation (6.22) shows that a displacement-dependent nonlinear tuning term needs to be considered for both frequency tuning and quadrature tuning when the drive amplitude is large. The nonlinear frequency terms will lead to a A-f effect as discussed in previous chapters, whereas the nonlinear quadrature tuning term will lead to a "dip" in the indirect (cross-coupling) frequency response as shown in Fig. 6.11. When quadrature is fully cancelled at the drive-mode resonant frequency, quadrature residuals will appear at other frequencies and causing higher bias level. The larger the drive amplitude, the deeper and narrower the dip will be.

Both the frequency and quadrature nonlinear tuning will lead to additional demodulation phase errors with drive amplitude variations, causing higher instability for larger drive amplitude and putting a trade-off between the achievable resolution and bias stability. Therefore, design modifications should be made to reduce the nonlinear tuning effect. As seen in (6.22), the nonlinear tuning term is proportional to the modal displacement coefficient α_1. Consequently, increasing the offset of the electrodes from the antinodes of the drive mode will reduce the nonlinearity, and the compromise in tuning efficiency can be overcome with larger electrode area or

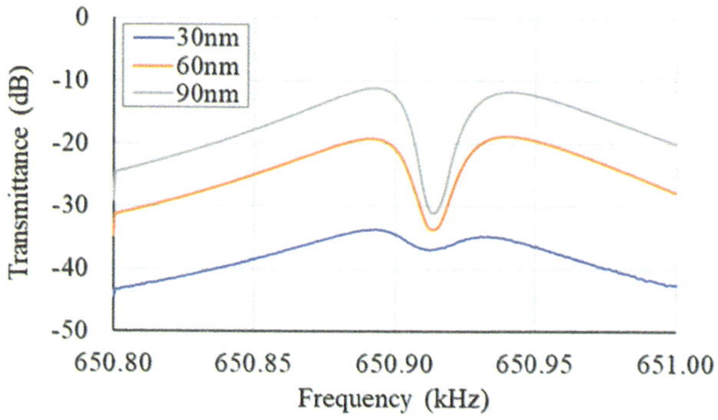

Fig. 6.11 The observed nonlinear quadrature tuning behaviors in the indirect channel for different drive amplitudes

higher tuning voltages. Another way to reduce the nonlinearity is by adding additional slanted electrodes that undergo opposite gap-size changings in both drive and sense modes as the tuning electrodes to be used. Such an electrode will generate the same linear tuning effect as the original electrode, but cancels the first-order nonlinear tuning term, reducing the nonlinearity without compromising the tuning efficiency. However, the increased number of electrodes may involve larger area and requires careful designs.

Chapter 7
Conclusions and Future Work

This dissertation presents a comprehensive study of physical phenomena governing the behaviors of generalized non-ideal MEMS resonant gyroscopes including pitch and roll gyroscopes involving out-of-plane degrees-of-freedom. Design solutions are developed and experimentally verified addressing the most challenging problems in multi-DOF gyroscopes performance scaling, which, for the first time, enabled a high-performance TIMU prototype with robust 3-axis resonant gyroscopes reaching tactical-grade and suitable for short-term navigational applications. A fundamental study of the performance limits in the high-frequency resonant gyroscopes is presented, pointing directions for future designs to reach higher performance toward MEMS-based robust INS.

7.1 Contributions

The major technical contributions made throughout this work are summarized as follows:

I. Performance scaling of resonant pitch and roll gyroscopes with multi-DOF:

   A. For the first time, a method for efficient quadrature cancellation in pitch and roll gyroscopes was developed and experimentally verified through the first anisotropic-wet-etching-based bulk-micromachined silicon resonant gyroscope.
   B. The incorporation of slanted electrodes into substrate-decoupled high-frequency resonant pitch and roll gyroscope designs was presented. And a novel HARPSS+ process was developed and verified to enable the implementation of such designs. The HARPSS+ process is by far one of the most complex and enabling MEMS fabrication platforms.

H. Wen, *Toward Inertial-Navigation-on-Chip*, Springer Theses,
https://doi.org/10.1007/978-3-030-25470-4_7

C. A novel method for creating silicon nano-gap comb-drives was proposed and successfully demonstrated, allowing enhancements in the gyroscope SF while reducing undesired nonlinearity effects.
D. A novel robust resonant framed-annulus pitch or roll gyroscope design with high Coriolis sensitivity was proposed and experimentally validated showing the highest Coriolis coupling coefficient among all high-frequency pitch or roll gyroscopes, and for the first time, high performance comparable with state-of-the-art BAW yaw gyroscopes.

II. High-performance single-chip TIMU:

A. A novel quasi-solid disk BAW yaw gyroscope was designed with large *Q*-factors and ultrahigh shock and vibration resistance for enabling robust IMUs.
B. For the first time, compact WLP single-chip TIMUs with 3-axis resonant gyroscopes were designed and fabricated with high yield, utilizing the quasi-solid disk BAW yaw gyroscope and high-frequency resonant framed-annulus pitch or roll gyroscopes. High-performance was demonstrated from all components on the TIMU, meeting requirements for short-term navigational applications.

III. Fundamental performance limits in high-frequency resonant gyroscopes:

A. Experimental investigations and numerical modeling were performed to identify the root of performance limits and, for the first time, provide quantitative evaluation of the behavior and contribution of each limiting factors. Strategies to address different limiting factors for further performance enhancement were discussed.
B. Differences between the performance scaling of yaw and pitch or roll resonant gyroscopes were presented, and additional source of errors in pitch or roll gyroscopes was identified with possible solutions discussed.

7.2 Future Work

The presented prototype resonant gyroscopes showed promising performance, opening opportunities for novel applications that cannot be achieved with conventional MEMS gyroscopes. The prototypes also served as good vehicles for resonant gyroscope performance limit investigation, which provided new understandings for future performance enhancements in high-frequency MEMS resonant gyroscopes.

7.2.1 Low-Power Resonant Gyroscopes

The concept of Internet of Things (IoT) is developing rapidly in recent years. Many emerging IoT applications such like personal health informatics and remote user interactions ask for integrated robust inertial sensors. Such consumer-oriented IoT applications require portable inertial sensors with small size, low cost, very low power, and high environmental robustness. While conventional MEMS gyroscopes show great advantages on the size and cost aspects, the power consumption is still a concern for long-standby-time close-to-zero-power applications, and their environmental robustness is not yet satisfying.

High-frequency resonant gyroscopes offer a clear advantage over their low-frequency counterparts in terms of environmental robustness. In addition, they also offer opportunities for ultra-low power operation with nanoscale transduction gaps. Conventional MEMS gyroscopes use mode-split operation with small mechanical SF. As a result, they require high-power electronics with enough energy injected to suppress electrical noise and achieve reasonable signal to noise ratios.

In comparison, resonant gyroscopes offer high sensitivity through sense mode mechanical Q-amplification, therefore highly reduce the power requirements in the interface electronics. What is more, the high-frequency resonant gyroscopes can provide high bandwidth and fast response time in both drive and sense modes while maintaining high quality-factors, which allows the implementation of less than 100% duty cycle operation and significantly reduces the effective power consumption. For example, the resonant framed-annulus pitch or roll gyroscope has a high drive mode Q of 13k at a high-operational frequency of ~660 Hz, which corresponds to a time constant of ~6.27 ms. The time constant is verified with a ring-down measurement on the mode-matched gyroscope as shown in Fig. 7.1. For a 50% duty cycle operation, the fast turn-on response of the gyroscope would still allow a bandwidth above 10 Hz, which is sufficient for most personal applications.

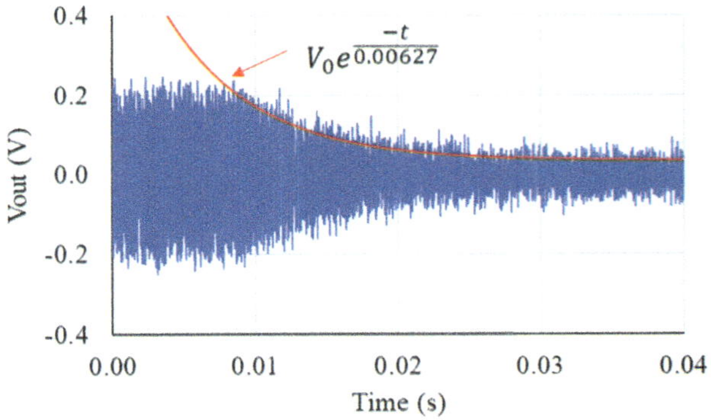

Fig. 7.1 Ring-down measurement of the resonant framed-annulus pitch or roll gyroscope

7.2.2 *Performance Enhancement in High-Frequency Resonant Gyroscopes*

As shown in Chap. 6, the long-term stability of high-frequency resonant gyroscopes is limited by environmental variation-induced drift in dissipation levels and interface electronics once the flicker noise limit is significantly reduced through proper quadrature cancellation.

7.2.2.1 Chip-Level Ovenization

Major environmental variations include ambient temperature variations, substrate stress variations, and random vibrations. With substrate-decoupled high-frequency designs, high stress and vibration robustness can be achieved, leaving temperature variation as the major environmental effect. As mentioned in previous chapters, a properly designed resonant gyroscope usually has TED limited quality-factors with well-defined temperature dependence. When the ambient temperature varies, a corresponding Q variation will lead to a bias level variation. In addition, for pitch and roll gyroscopes using nondegenerate modes, the small difference in TCFs of the drive and sense modes will lead to relative frequency shift with temperature variations, which affects the demodulation phase and results in bias drift (RRW or RR).

The prototype TIMU developed in the work has on-chip resistive heaters integrated in the same vacuum cavity as the 3-axis gyroscopes. This provides the possibility of using chip-level ovenization to achieve temperature-insensitive stable gyroscopes. The efficiency of the heaters for heating the TIMU chip is examined by monitoring the pitch gyroscope resonant frequencies when passing different currents through the heaters [1]. Measured results are shown in Fig. 7.2. The results indicate efficient heating of the gyroscope cavity up to above 100 °C with a power less than 600 mW. When combined with the timing resonator, a closed-loop temperature control will be possible, allowing the vacuum cavity of the TIMU to be ovenized at ~90 °C and remain stable with operational ambient temperatures ranging from −40 °C to 85 °C.

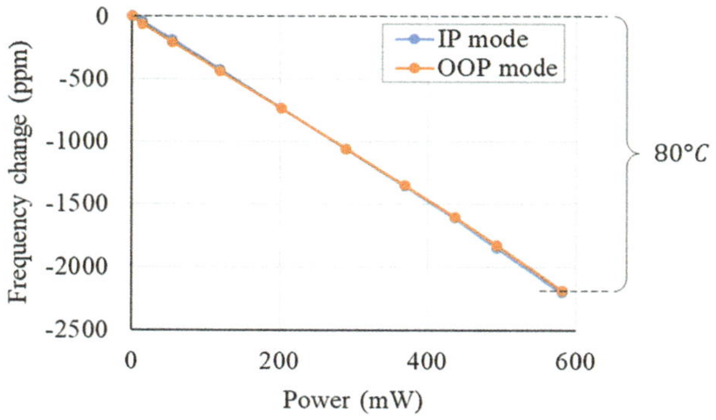

Fig. 7.2 Measured pitch gyroscope operational modes frequency shift with different heater powers [1]

7.2.2.2 System Optimization and ASIC Implementation

The high-frequency resonant gyroscopes presented in this work so far have only been characterized with discrete front-end electronics and simple drive mode control loops (PLL + AGC) implemented using a HF2LI lock-in amplifier. No dynamic tuning controls have been incorporated to compensate for variations in frequency mismatch and cross-coupling. Large variations over time and environmental conditions as well as due to inaccuracies in initial tuning setup may cause deviations from a close-to-mode-matched operation and subject the gyroscope to potential reliability and performance repeatability issues. To overcome such issues, closed-loop frequency tuning and quadrature cancellation control can be incorporated in the system to realize in-run automatic mode-matching and enhance the gyroscope performance consistency and reliability [2–4].

In the current system, discrete electronics and instruments are exposed to the ambient and connected through long cables, which presents phase and offset errors dominating the bias instability of the gyroscope. Therefore, it is desirable to develop low-noise application-specific integrated circuit (ASIC) for high-frequency gyroscope interfacing [5]. The ASIC and MEMS units can be integrated through flip-chip technology, which will not only reduce the form-factor and power consumption but also provide better isolation of the system from environmental variations, therefore improving the gyroscope stability.

7.2.2.3 Quality-Factor Optimization

Once the bias instability and long-term bias drift limits are eliminated or properly compensated, the ADEV will keep reducing with increased integration time following the $-1/2$ slope line defined by the mechanical white noise limit. As given in (2.57), this limit depends on the quality-factor of the gyroscope, which can put an ultimate limit on the achievable resolution of a MEMS gyroscope. Therefore, improving the quality-factor of the gyroscope is very important in realizing inertial-grade MEMS gyroscopes.

As mentioned in Chap. 3, the major quality-factor limits in high-frequency gyroscopes are TED, SFD, anchor loss, and Akheiser limit. In current designs, anchor loss is addressed by implementing substrate-decoupled designs, and SFD can be minimized by using higher-frequency designs as well as improving the cavity pressure through more advanced capping process such as incorporate getters. As a result, most high-frequency designs are currently limited by TED, which prevents them from reaching the ultimate Akheiser limit of SCS with the fQ-product of ~2×10^{13}. To further reduce TED, solid structures without thin beams and release holes may be used. For example, fQ-product ~5×10^{12} has been demonstrated for a fourth-order wine-glass mode of a nearly solid disk processed by backside releasing with only decoupling notches and no release holes, which is close to the Akheiser limit [6]. With further improvements in the resonator design and fabrication process, higher *fQ*-product may become possible even for the lower-order gyroscopic modes, which can ultimately offer quality-factors in the order of 10M, providing more than ten times better resolution than the limit of current designs and reaching inertial-grade performance.

References

1. H. Wen, A. Daruwalla, Y. Jeong, P. Gupta, J. Choi, C.S. Liu, F. Ayazi, Wafer-level-packaged HARPSS+ MEMS platform: Integration of robust timing and inertial measurement units (TIMU) on a single chip, in *Proceedings of 2018 IEEE/ION Position, Location and Navigation Symposium (PLANS 2018)*, (2018), pp. 261–266. © 2018 IEEE. Figures and tables reprinted with permission
2. A. Sharma, M.F. Zaman, F. Ayazi, A 0.2°/hr micro-gyroscope with automatic CMOS mode matching, in *2007 IEEE International Solid-State Circuits Conference. Digest of Technical Papers*, (2007), pp. 386–610
3. I.P. Prikhodko et al., Mode-matched MEMS Coriolis vibratory gyroscopes: Myth or reality? in *2016 IEEE/ION Position, Location and Navigation Symposium (PLANS)*, (2016), pp. 1–4
4. I.P. Prikhodko et al., Half-a-month stable 0.2 degree-per-hour mode-matched MEMS gyroscope, in *2017 IEEE International Symposium on Inertial Sensors and Systems (INERTIAL)*, (2017), pp. 1–4
5. D.E. Serrano et al., Environmentally-robust high-performance tri-axial bulk acoustic wave gyroscopes, in *2016 IEEE/ION Position, Location and Navigation Symposium (PLANS)*, (2016), pp. 5–8
6. R. Mirjalili et al., Substrate-decoupled silicon disk resonators having degenerate gyroscopic modes with Q in excess of 1-million, in *2015 Transducers – 2015 18th International Conference on Solid-State Sensors, Actuators and Microsystems (TRANSDUCERS)*, (2015), pp. 15–18

Appendix A: Fabrication Process

This appendix shows the detailed flows of the microfabrication processes developed in this work. The color code for materials used in the processes is listed in Fig. A.1.

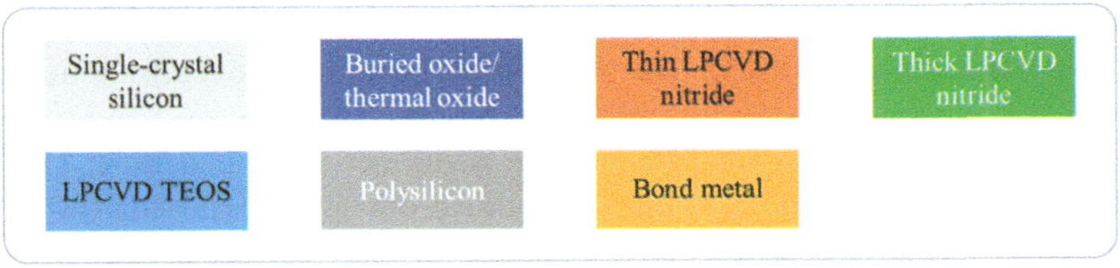

Fig. A.1 Color code for processed materials

H. Wen, *Toward Inertial-Navigation-on-Chip*, Springer Theses,
https://doi.org/10.1007/978-3-030-25470-4

A.1 Wet-Etching-Based Bulk Micromachining Process

The anisotropic wet-etching-based bulk micromachining process does not involve expensive high-aspect-ratio DRIE. The process creates bulk resonant structures with isosceles trapezoidal cross sections and electrodes on top and slanted surfaces with nanoscale capacitive transduction gaps [1].

The process starts with two-mask wet-etching steps [2] on a (100) SOI wafer, considering the quad-mass pitch or roll gyroscopes presented in Chap. 3 as an exemplary device. First, a (100) SOI wafer with 40 μm device layer is annealed at 1100 °C to enhance the BOX quality. A 350-nm-thick low-stress LPCVD silicon nitride mask layer (thick nitride) is deposited on the SOI wafer. The thick nitride is patterned with the device trench patterns aligned to (110) crystal directions using photolithography and RIE etching. A second 150 nm thick low-stress LPVCD nitride mask layer (thin nitride) is deposited after chemical cleaning of the wafer. The thin nitride is patterned using timed RIE etching with self-aligning openings to partially expose the trench patterns while covering regions near the convex corners where trenches perpendicular to each other meet.

After the thin nitride is patterned, an anisotropic silicon wet-etching is performed in a 65 °C 45% KOH bath for around 2 h. The exposed (100) surface is etched down and (111) surfaces are created. The wet-etching undercuts the self-aligning patterns and results in trench patterns defined solely by the thick nitride patterns. The wet-etching is self-confined, but etching time should not be excessively longer than needed to reduce the slow etching of BOX layer and (111) planes (Fig. A.2).

The wafer is thermally oxidized to form 500 nm of local oxide on the (111) planes using wet oxidation at 1100 °C. The thin nitride layer is removed by a time blank RIE, and the remaining trench patterns are exposed (Fig. A.3).

A second anisotropic silicon wet-etching is performed again in a 65 °C 45% KOH bath for around 2 h. The wet-etching etches down the newly exposed (100) surface and undercuts the silicon at the convex corners of the local oxide layer. Once the wet-etching is complete, the resonant structures along with electrode pads are completely formed. The local oxide and nitride layers are then removed together using blank RIE both on the front and back of the wafer, and the wafer is cleaned and prepared for the next steps (Fig. A.4).

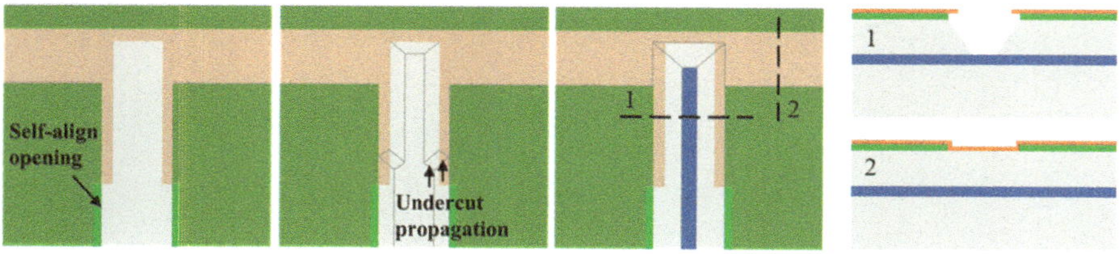

Fig. A.2 Top views of the two-layer self-aligned nitride masks and etching trend of the first silicon anisotropic wet-etching and cross-sectional views of the resulted structures [1]

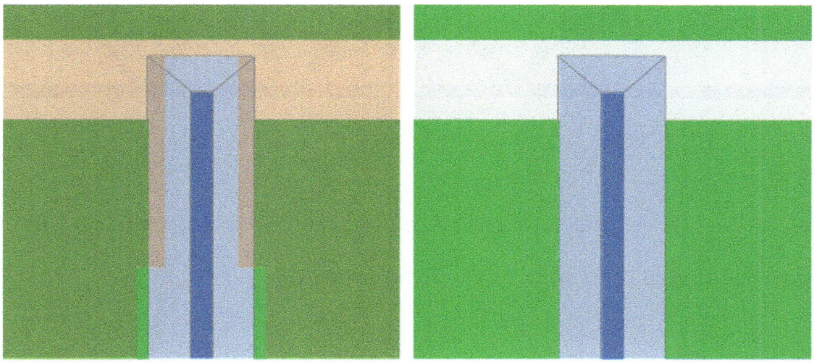

Fig. A.3 Local oxidation of (111) silicon planes and thin nitride mask layer removal [1]

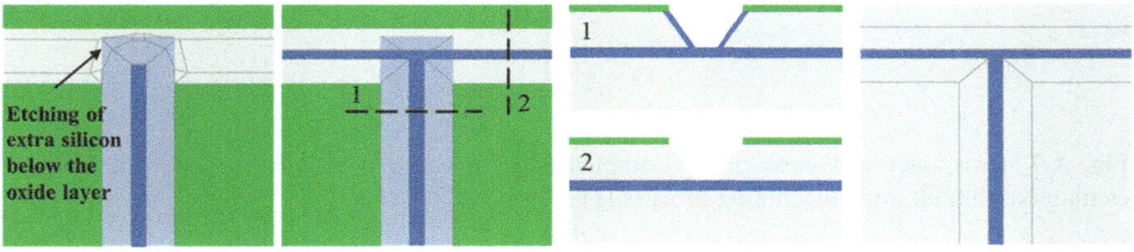

Fig. A.4 Top views of the second wet-etching procedure and cross-sectional views of the resulted structures [1]

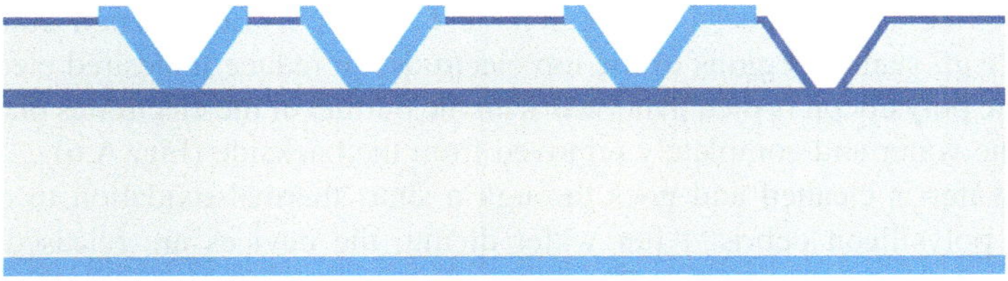

Fig. A.5 Thick LPCVD TEOS deposition, patterning, and thin sacrificial thermal oxide formation [1]

A 3-μm-thick LPCVD TEOS layer is conformally deposited on the wafer and selectively patterned to expose the surfaces for the formation of nanoscale transduction gaps and anchor points to the electrode pads. The anisotropically wet-etched wafer presents high topography, and proper lithography is achieved using spray-coated photoresist instead of conventional spin-coated photoresist. After TEOS patterning, the wafer is cleaned and thermally oxidized with dry oxidation at 1100 °C to form sacrificial oxide that defines the capacitive transduction gap-size. For the exemplary device, 510 nm and 540 nm thermal oxide is formed on the top and slanted silicon surfaces, respectively. The slanted oxide layer is slightly thicker due to the slightly faster oxidation of (111) plane than (100) plane (Fig. A.5).

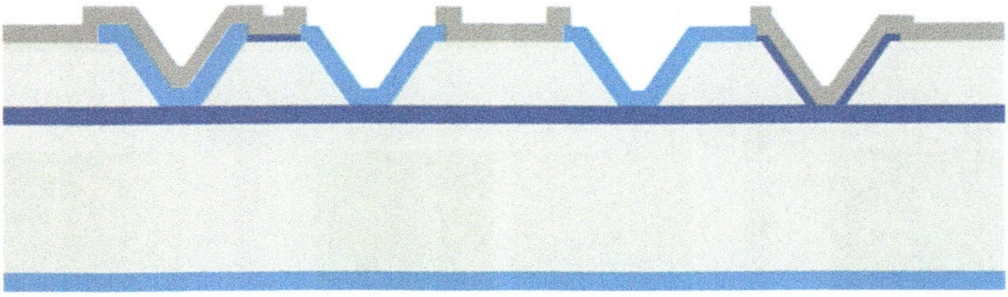

Fig. A.6 Top and slanted polysilicon electrode formation [1]

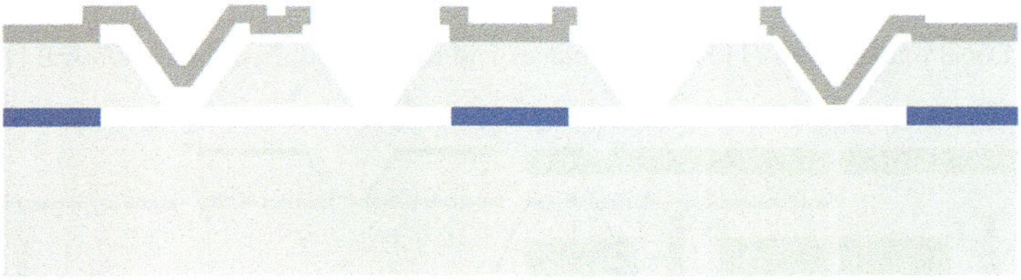

Fig. A.7 Cross-sectional view of the completed and released device fabricated with the wet-etching-based bulk micromachining process [1]

The thin thermal oxide is then patterned and etched off from the electrode pads for electrode anchoring. A 7.5 μm polysilicon layer is formed through a few repeated LPCVD deposition and annealing steps. The polysilicon is directly attached to the SCS electrode pads and is separated from the SCS resonator by the nanoscale thermal oxide at the transduction regions and by thick TEOS at non-transduction regions (e.g., slanted regions of the top electrode) to reduce undesired electrostatic force. The polysilicon is then patterned with the outline of the electrodes on the front side of the wafer and completely removed from the backside (Fig. A.6).

The wafer is cleaned and goes through a short thermal oxidation to eliminate possible polysilicon debris. After wafer dicing, the devices are released in 49% hydrofluoric acid (HF) solution. The released devices are immersed in isopropanol (IPA) and dried using a supercritical point dryer to avoid stiction risks (Fig. A.7).

A.2 HARPSS+ Process

The high-aspect-ratio poly- and single-crystal silicon RIE plus wet-etching (HARPSS+) process developed in this work is one of the most complicated and versatile MEMS batch fabrication platforms to date [3]. The HARPSS+ process enables nano-gap electrodes in vertical, horizontal, and slanted orientations; therefore it is suitable for high-yield batch fabrication of integrated MEMS sensor units with wafer-level packaged robust high-frequency components.

Taking the TIMU presented in this work as an example, robust devices are fabricated on a (100) SOI wafer with 40 μm device layer. It is worth noting that the HARPSS+ process is not limited to the specific device thickness. As a matter of fact, the annulus gyroscope with slanted electrodes which is also presented in this work in Chap. 3 was implemented using HARPSS+ process on a 60 μm SOI wafer. For illustration purpose, the process flow is detailed as follows with specifications given for the batch with 40 μm device thickness.

First, the SOI wafer is wet-oxidized at 1100 °C to form a 3 μm oxide mask layer. The oxide is patterned with device outlines, and silicon DRIE is performed to create 4.5-μm-wide vertical trenches through the device layer. The wafer is then cleaned and dry-oxidized at 1100 °C to form 270 nm sacrificial oxide on all the vertical sidewalls. The trenches are conformally filled with LPCVD polysilicon through repeated deposition, annealing, and etching-back steps. Once the trenches are completely filled, polysilicon is etched back with dry etching on both front and back of the wafer. The polysilicon inside the trenches is selectively etched using DRIE, leaving polysilicon only at the regions where vertical electrodes are to be formed (Fig. A.8).

Second, the wafer is cleaned and shortly oxidized to eliminate polysilicon debris, and the trenches are refilled with LPCVD TEOS through repeated deposition, annealing, and etching-back, leaving ~3 μm oxide on the front of the wafer by the end of the filling procedures. The TEOS is patterned to expose top transduction and electrode anchoring regions. The wafer is then dry-oxidized at 1100 °C for around 75 nm oxide to partially form the top sacrificial oxide layer (Fig. A.9).

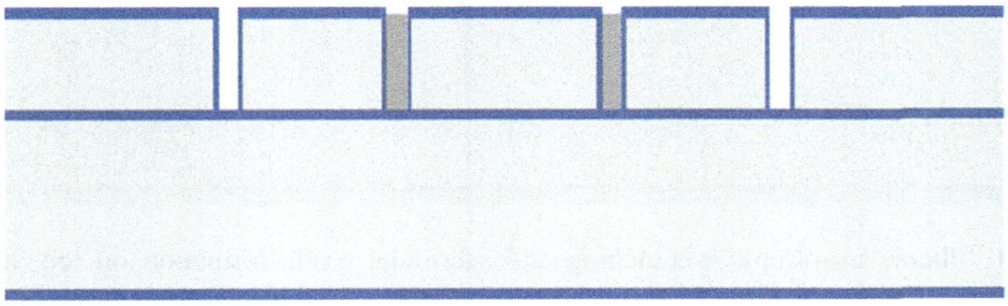

Fig. A.8 DRIE trench formation and trench polysilicon selectively etching with sacrificial oxide formed on the vertical sidewalls [3]

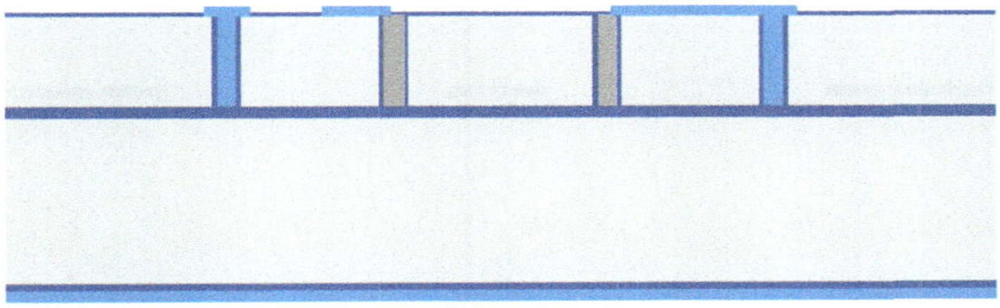

Fig. A.9 TEOS trench refilling and partial top sacrificial oxide formation [3]

Thirdly, TEOS on the frontside of the wafer is patterned with openings for anisotropic wet-etching. The wet-etching openings along with trench patterns nearby are specially design as described previously in Chap. 3. The silicon anisotropic wet-etching is performed in an 85 °C 25% TMAH bath for about 1.5 h after dipping in the buffered oxide etchant for a few seconds to remove the native oxide. After the wet-etching, the wafer is dry-oxidized at 1100 °C to completely form sacrificial oxide on top and slanted surfaces with a final thickness of 270 nm for both. The top sacrificial oxide is then etched away on the electrode pads for electrode anchoring (Fig. A.10).

Lastly, about 7 μm LPCVD polysilicon is conformally deposited and patterned, forming the top and slanted electrodes, as well as connecting the trench polysilicon to form vertical electrodes. After the polysilicon etching, the whole wafer is shortly oxidized and released in 49% HF solution. Again, supercritical drying can be used to dry the released wafer without stictions, and the completed device wafer is ready to be wafer-level packaged by bonding to a silicon capping wafer (Fig. A.11).

The capping wafer is fabricated using a regular silicon wafer. Deep via holes are first etched into the wafer with oxide mask. The wafer is then oxidized to form an electrical isolation layer. The holes are filled with LPCVD polysilicon which is etched back to the oxide surface by the end of the filling. Oxidation and patterning steps are followed to enclose the polysilicon leaving only small openings on top of the polysilicon. A layer of gold is evaporated and patterned as bond metal. Deep

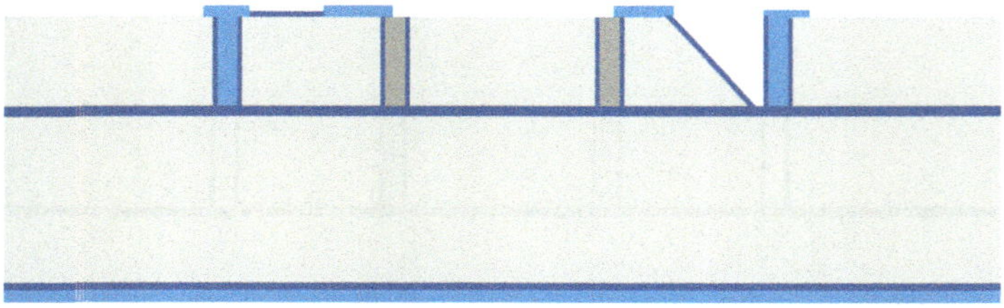

Fig. A.10 Silicon anisotropic wet-etching and sacrificial oxide formation on top and slanted surfaces [3]

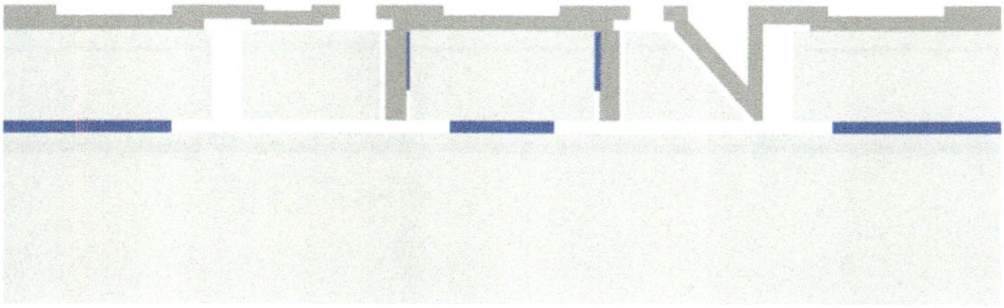

Fig. A.11 Cross-sectional view of released device wafer with nano-gap vertical, horizontal, and slanted polysilicon electrodes [3]

Fig. A.12 Pre-bonding preparation of the capping wafer [3]

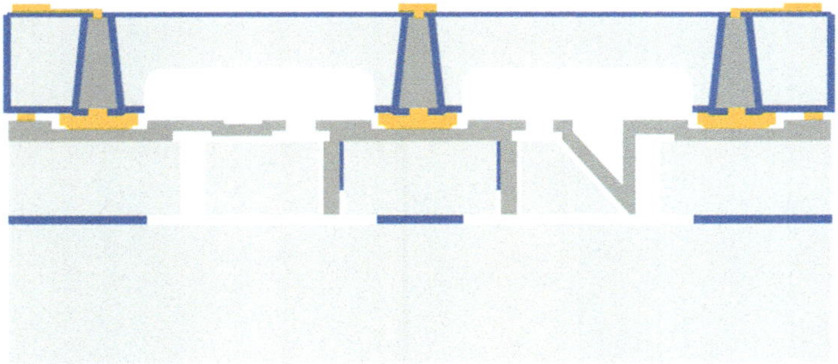

Fig. A.13 The final cross-sectional view of WLP HARPSS+ processed devices [3]

cavities are then etched into the wafer with patterns matching the footprint of the device wafer (Fig. A.12).

The capping wafer is then bonded to the device wafer through vacuum eutectic bonding. The capping wafer is thinned down after bonding to expose polysilicon vias. A layer of PECVD oxide is then deposited and pattern for electrical isolations. Metal traces and pads are formed at last through evaporation and electroplating (Fig. A.13).

References

1. H. Wen, A. Daruwalla, F. Ayazi, Resonant pitch and roll silicon gyroscopes with sub-micron-gap slanted electrodes: breaking the barrier toward high-performance monolithic inertial measurement units, Microsyst. Nanoeng. **3**, 16092 (2017)
2. P. Prem, C. Sudhir, A novel process for perfect convex corner realization in bulk micromachining, J. Micromech. Microeng. **14**(10), 1416 (2004)
3. H. Wen, A. Daruwalla, Y. Jeong, P. Gupta, J. Choi, C.S. Liu, F. Ayazi, Wafer-level-packaged HARPSS+ MEMS platform: integration of robust timing and inertial measurement units (TIMU) on a single chip, in *Proceedings of 2018 IEEE/ION Position, Location and Navigation Symposium (PLANS 2018)*, (IEEE, 2018), pp. 261–266. © 2018 IEEE. Figures and tables reprinted with permission

Appendix B: Numerical Models

This appendix lists the approximation method used for estimating the mechanical nonlinearity in high-frequency annulus resonators and the MATLAB code used for numerical modeling of the bias stability in high-frequency resonant gyroscopes.

B.1 Annulus Resonator Mechanical Nonlinearity Estimation

Mechanical nonlinearity of SCS resonators depends on the doping conditions of the substrate and the mode shape under consideration. Theoretical and experimental studies on mechanical nonlinearities in high-frequency SCS resonators have been widely reported but mostly in the context of length extension (LE) mode resonators [1–3]. References on the $n = 2$ wineglass mode of SCS annulus resonator are very limited due to its complicated strain distribution.

An equivalent LE model with scaled nonlinear Young's modulus is introduced in this work to qualitatively estimate the mechanical nonlinearity of the $n = 2$ wineglass mode [4]. The equivalent model is constructed by dividing the annulus into small nonlinear spring-mass elements and mapping them to the shape of a LE resonator (Fig. B.1).

In the approximated model, contributions to nonlinearity form the s`mall radial and shear stress in the annulus are neglected. The Lagrangian of the infinitesimal element system can then be written as:

H. Wen, *Toward Inertial-Navigation-on-Chip*, Springer Theses,
https://doi.org/10.1007/978-3-030-25470-4

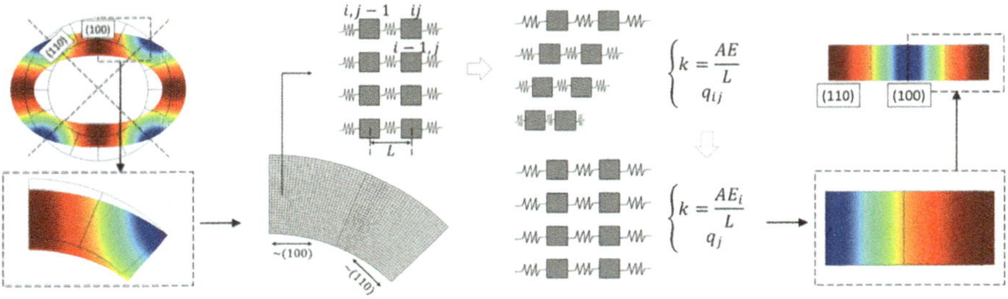

Fig. B.1 Equivalent LE model of the $n = 2$ mode with the annulus divided into infinitesimal elements. The annulus is mapped into a LE model by transferring the i-distribution of strain from q to k [4]

$$\begin{aligned}\mathcal{L} &= \sum_{ij} (T_{ij} - U_{ij}) = \sum_{ij} \left[\frac{1}{2} L_i A \rho \dot{\boldsymbol{q}}_{ij}^2 - \sum_n \frac{k_n}{n+2} \left(\boldsymbol{q}_{ij} - \boldsymbol{q}_{i,j-1} \right)^{n+2} \right] \\ &= \sum_{ij} L_i A \left[\frac{1}{2} \rho \dot{q}_{ij}^2 - \sum_n \frac{E_n}{n+2} \left(\frac{\Delta L_{ij}}{L_i} \right)^{n+2} \right] \\ &= \sum_{ij} L_i A \left[\frac{1}{2} \rho \dot{q}_{ij}^2 - \sum_n \frac{E_n}{n+2} S_{ij}^{n+2} \right], \end{aligned} \tag{B.1}$$

where $\boldsymbol{q}_{ij}$, S_{ij}, L_i, A, and E_n are the displacement vector, strain, length, cross-sectional area, and nth order Young's modulus of element ij. Lagrangian mechanics gives:

$$\frac{d}{dt}\left(\frac{\partial \mathcal{L}}{\partial \dot{q}} \right) - \frac{\partial \mathcal{L}}{\partial q} = \sum_{ij} \left[\frac{d}{dt} \left(\rho \dot{q}_{ij}^2 \frac{\partial \dot{q}_{ij}}{\partial \dot{q}} \right) + \sum_n E_n S_{ij}^{n+1} \frac{\partial S_{ij}}{\partial q} \right] = 0. \tag{B.2}$$

Dividing the annulus into ~(100) and ~(110) sections and assuming constant E_n in each section, the i (width direction) and j (arc direction) strain distribution can be decoupled as below:

$$\begin{aligned} &\sum_{ij} \left[\frac{d}{dt} \left(c_{1ij} \rho \dot{q} \frac{\partial \dot{q}}{\partial \dot{q}} \right) + \sum_n E_n (f_i S_j)^{n+1} f_i \frac{\partial S_j}{\partial q} \right] \\ &= c_1 \rho \ddot{q} + \sum_{ij} \left[\sum_n \left(E_n f_i^{n+2} \right) S_j^{n+1} \frac{\partial S_j}{\partial q} \right] \\ &= c_1 \rho \ddot{q} + \sum_j \left[\sum_n (a_n E_n) S_j^{n+1} \frac{\partial S_j}{\partial q} \right] \\ &= c_1 \rho \ddot{q} + \sum_j E' S_j \frac{\partial S_j}{\partial q}, \end{aligned} \tag{B.3}$$

Table B.1 Estimated A-*f* coefficient for $n = 2$ wineglass mode [4]

Dopant/doping level	Simulated A-*f* coefficient
NP 1.78 mΩ·cm	7.44×10^{13} Hz/m^2
NA 3.1 mΩ·cm	6.23×10^{13} Hz/m^2
NS 17.1 mΩ·cm	5.39×10^{13} Hz/m^2

where f_i is the normalized i-distribution of strain, S_j is i-index averaged j-distribution of strain, and c_1 and a_n are integration constants. The value of a_n can be evaluated as:

$$a_n = \int_0^{W_{\text{annulus}}} \frac{[f(w)]^{n+2}}{W_{\text{annulus}}} dw, \tag{B.4}$$

with the distribution function $f(w)$ obtained numerically using FEM simulation. Substituting (B.3) into (B.2) gives:

$$c_1\rho\ddot{q} + \sum_j E' S_j \frac{\partial S_j}{\partial q} = c_1\rho\ddot{q} + c_2\sum_j \frac{1}{2} k_j' \frac{\partial (q_j - q_{j-1})^2}{\partial q} = 0. \tag{B.5}$$

Since S_j in the $n = 2$, wineglass mode resembles the length strain distribution in an LE mode. Eq. (B.5) can be considered as an LE resonator with a scaled nonlinear Young's modulus E', causing the same nonlinearity level as the annulus resonator with the same maximum strain. The resonant frequency and strain-displacement relation can be matched between the annulus and the equivalent LE model by adjusting the length constant (c_2) and density constant (c_1) of the LE element in FEM simulation, which allows the equivalent model to produce an approximation of the A-*f* coefficient κ through lumped-element numerical simulations. Based on SCS material properties extracted from available experimental data for LE resonators [1, 2], such numerical simulations are conducted to estimate mechanical nonlinearity of the annulus $n = 2$ wineglass mode implemented in substrates with different doping conditions, and the results are listed in Table B.1.

B.2 MATLAB Code for Numerical Bias Instability Modeling

```
%% ----Define functions----
syms m k1 k2 d1 d2 lambda w0 Fd ome thetaw thetat CFT

% Drive displacement
Drive=(Fd*((sin(thetaw))^2*k1+(cos(thetaw))^2*k2+1i*w0*((sin
(thetat))^2*d1+(cos(thetat))^2*d2+1i*m*w0)))/(k1*k2+1i*(d1*((sin
(thetat-thetaw))^2*k1+(cos(thetat-thetaw))^2*k2)+d2*((cos(thetat-
thetaw))^2*k1+(sin(thetat-thetaw))^2*k2))*w0-(d1*d2+m*(k1+k2))*
w0^2-1i*m*(d1+d2)*w0^3+m^2*w0^4);
% Bias displacement
Offset=CFT*Fd*w0+1/2*(Fd*(-1i*sin(2*thetat)*(d1-d2)*w0+sin
```

```
(2*thetaw)*(-k1+k2)))/(k1*k2+1i*(d1*((sin(thetat-thetaw))^2*k1
+(cos(thetat-thetaw))^2*k2)+d2*((cos(thetat-thetaw))^2*k1+(sin
(thetat-thetaw))^2*k2))*w0-(d1*d2+m*(k1+k2))*w0^2-1i*m*(d1+d2)*w0^3
+m^2*w0^4);
% Coriolis displacement for 1dps
Rate=(2i*lambda*Fd*m*w0*pi/180)/(k1*k2+1i*(d1*((sin(thetat-thetaw))
^2*k1+(cos(thetat-thetaw))^2*k2)+d2*((cos(thetat-thetaw))^2*k1
+(sin(thetat-thetaw))^2*k2))*w0-(d1*d2+m*(k1+k2))*w0^2-1i*m*(d1
+d2)*w0^3+m^2*w0^4);

DV=matlabFunction(Drive); %@(Fd,d1,d2,k1,k2,m,thetat,thetaw,w0)
SF=matlabFunction(Rate); %@(Fd,d1,d2,k1,k2,lambda,m,thetat,thetaw,
w0)
OS=matlabFunction(Offset);  %@(CFT,Fd,d1,d2,k1,k2,m,thetat,thetaw,
w0)
projection=@(r1,t1,t2) vpa(r1*cos(t1-t2));
rejection=@(r1,t1,t2) vpa(r1*sin(t1-t2));

%% ----Initial conditions----
digits(16); % Numerical resolution

Fd=vpa(1e5); % Drive force
CFT=vpa(0.0); % Feedthrough (equivalent)

m=vpa(15e-9); % Effective mass in kg
lambda=vpa(0.6); % Coriolis coupling coefficient

w11=vpa(2*pi*4.25e6); % Pseudo drive mode frequency
w22=vpa(w11+0.*2*pi); % Pseudo sense mode frequency

Q1=vpa(1e5+50); % Drive mode Q
Q2=vpa(1e5-50); % Sense mode Q

k11=vpa(m*w11^2);
k22=vpa(m*w22^2);
k12=-k11*0.000000425*8/8; % Stiffness-coupling to match measured
quadrature
d11=vpa(m*w11/Q1);
d22=vpa(m*w22/Q2);
d12=d11*0.0058; % Damping-coupling to match measured in-phase bias

thetaw=vpa(angle(((k11-k22)+1i*2*k12))/2); % Principle stiffness
axis misalignment
thetat=vpa(angle(((d11-d22)+1i*2*d12))/2); % Principle damping axis
misalignment

% Calculate principle stiffness and damping constant
if (k11-k22)~=0
  k1=vpa((k11+k22)/2+(k11-k22)/(2*cos(2*thetaw)));
  k2=vpa((k11+k22)/2-(k11-k22)/(2*cos(2*thetaw)));
else
  if k12~=0
```

```
    k1=vpa((k11+k22)/2+k12/(sin(2*thetaw)));
    k2=vpa((k11+k22)/2-k12/(sin(2*thetaw)));
  else
    k1=k11;
    k2=k11;
  end
end

if (d11-d22)~=0
  d1=vpa((d11+d22)/2+(d11-d22)/(2*cos(2*thetat)));
  d2=vpa((d11+d22)/2-(d11-d22)/(2*cos(2*thetat)));
else
  if d12~=0
    d1=vpa((d11+d22)/2+d12/(sin(2*thetat)));
    d2=vpa((d11+d22)/2-d12/(sin(2*thetat)));
  else
    d1=d11;
    d2=d11;
  end
end

w0=vpa(sqrt(k1/m)); % Lock-in frequency (observed drive mode
frequency)

% PLL sweep setting
f_start=w0/(2*pi)-0.0008;
f_end=w0/(2*pi)+0.0008;
f_step=0.000016;
N_step=(f_end-f_start)/f_step;

%% ----Initial values----
DV0=DV(Fd,d1,d2,k1,k2,m,thetat,thetaw,w0);
ampDV0=vpa(abs(DV0));
phDV0=vpa(angle(DV0));

SF0=vpa((SF(Fd,d1,d2,k1,k2,lambda,m,thetat,thetaw,w0)));
ampSF0=vpa(abs(SF0));
phSF0=vpa(angle(SF0));

OS0=vpa(OS(CFT,Fd,d1,d2,k1,k2,m,thetat,thetaw,w0));
ampOS0=vpa(abs(OS0));
phOS0=vpa(angle(OS0));

IP0=projection(ampOS0,phOS0,phSF0);
QD0=rejection(ampOS0,phOS0,phSF0);

% Isolation level in dB
ISO=20*log10(2*ampOS0/ampDV0); % Factor of 2 for electrode area
difference
ISOIP=20*log10(2*IP0/ampDV0);
ISOQD=20*log10(2*QD0/ampDV0);
```

```
% Bias equivalent rate output in dph
BI0=vpa(IP0*3600/ampSF0);
BQ0=vpa(QD0*3600/ampSF0);

%% ----Frequency noise----
dw=vpa(2*pi*0.000072);
% dk1=vpa((2*sqrt(k1/m)+dw)*dw*m);
% dk2=0;
dk1=0;
dk2=vpa((2*sqrt(k2/m)+dw)*dw*m);
dphi_TIA=vpa(2.3e-6); % TIA phase shift

% PLL
dphDV0_dw=vpa(angle(DV(Fd,d1,d2,k1+dk1,k2+dk2,m,thetat,thetaw,
w0))-phDV0);
dphDVminn=0;

for n=1:N_step
  f(n)=f_start+(n-1)*f_step;
  phDV_dw_sweep(n)=vpa(angle(DV(Fd,d1,d2,k1+dk1,k2+dk2,m,thetat,
thetaw,2*pi*f(n))));
  dphDV_dw_sweep(n)=vpa(phDV_dw_sweep(n)-phDV0);
  phSF_dw_sweep(n)=vpa(angle(SF(Fd,d1,d2,k1+dk1,k2+dk2,lambda,m,
thetat,thetaw,2*pi*f(n))));
  dphSF_dw_sweep(n)=vpa(phSF_dw_sweep(n)-phSF0);
  if abs(dphDV_dw_sweep(n))<=abs(dphDV0_dw)
    dphDV0_dw=dphDV_dw_sweep(n);
    dphDVminn=n;
  end
end

% Updated lock-in frequency
if dphDVminn~=0
  w0_dw=vpa(f(dphDVminn)*2*pi);
else
  w0_dw=w0;
end

% Plot
set(gcf,'Position',[0,45,960,200]);
figure (1);
plot(f,dphDV_dw_sweep*180/pi,'LineWidth',2);
hold on;
plot(f,dphSF_dw_sweep*180/pi,'LineWidth',2);
hold off;
title('Drive and Coriolis displacement phase shift');

% Updated values
DV_dw=DV(Fd,d1,d2,k1+dk1,k2+dk2,m,thetat,thetaw,w0_dw);
ampDV_dw=vpa(abs(DV_dw));
phDV_dw=vpa(angle(DV_dw));
vpa(phDV_dw-phDV0);
```

```
OS_dw=vpa(OS(CFT,Fd,d1,d2,k1+dk1,k2+dk2,m,thetat,thetaw,w0_dw));
ampOS_dw=vpa(abs(OS_dw));
phOS_dw=vpa(angle(OS_dw));
dampOS=vpa((ampOS_dw-ampOS0)/ampSF0);
dphOS_ddw=vpa(phOS_dw-phOS0);

SF_dw=vpa((SF(Fd,d1,d2,k1+dk1,k2+dk2,lambda,m,thetat,thetaw,
w0_dw)));
phSF_dw=vpa(angle(SF_dw));
dphase_dw=vpa(phSF_dw-phSF0) % Total demodulation phase error
% BI_ph_dw=vpa(sqrt(2)*dphase_dw*BQ0); % BI due to demodulation error

IP_dw=projection(ampOS_dw,phOS_dw,phSF0);
QD_dw=rejection(ampOS_dw,phOS_dw,phSF0);

BI_dw=vpa(sqrt(2)*abs(BI0-IP_dw*3600/ampSF0)) % In-phase BI in dph,
sqrt(2) for uncorrelated k1 and k2 error
BQ_dw=vpa(sqrt(2)*abs(BQ0-QD_dw*3600/ampSF0)) % Quadrature BI in dph,
sqrt(2) for uncorrelated k1 and k2 error

%% ----PLL error----
dphDV0_PLL=vpa(dphi_TIA);
dphDVminn=0;

for n=1:N_step
  f(n)=f_start+(n-1)*f_step;
  phDV_PLL_sweep(n)=vpa(angle(DV(Fd,d1,d2,k1,k2,m,thetat,
thetaw,2*pi*f(n))));
  dphDV_PLL_sweep(n)=vpa(phDV_PLL_sweep(n)-phDV0+dphi_TIA);
  phSF_PLL_sweep(n)=vpa(angle(SF(Fd,d1,d2,k1,k2,lambda,m,thetat,
thetaw,2*pi*f(n))));
  dphSF_PLL_sweep(n)=vpa(phSF_PLL_sweep(n)-phSF0);
  if abs(dphDV_PLL_sweep(n))<=abs(dphDV0_PLL)
    dphDV0_PLL=dphDV_PLL_sweep(n);
    dphDVminn=n;
  end
end

% Updated lock-in frequency
if dphDVminn~=0
  w0_PLL=vpa(f(dphDVminn)*2*pi);
else
  w0_PLL=w0;
end

%% Plot
% set(gcf,'Position',[0,45,960,200]);
% figure (1);
% plot(f,dphDV_PLL_sweep*180/pi,'LineWidth',2);
% hold on;
% plot(f,dphSF_PLL_sweep*180/pi,'LineWidth',2);
```

```
% hold off;
% title('Drive and Coriolis displacement phase shift');

% Updated values
DV_PLL=DV(Fd,d1,d2,k1,k2,m,thetat,thetaw,w0_PLL);
ampDV_PLL=vpa(abs(DV_PLL));
phDV_PLL=vpa(angle(DV_PLL));
vpa(phDV_PLL-phDV0);

OS_PLL=vpa(OS(CFT,Fd,d1,d2,k1,k2,m,thetat,thetaw,w0_PLL));
ampOS_PLL=vpa(abs(OS_PLL));
phOS_PLL=vpa(angle(OS_PLL));
dampOS_PLL=vpa((ampOS_PLL-ampOS0)/ampSF0);
dphOS_PLL=vpa(phOS_PLL-phOS0);

SF_PLL=vpa((SF(Fd,d1,d2,k1,k2,lambda,m,thetat,thetaw,w0_PLL)));
phSF_PLL=vpa(angle(SF_PLL));
dphase_PLL=vpa(phSF_PLL-phSF0); % Demodulation phase error
% BI_ph_PLL=vpa(dphase_PLL*BQ0) % BI due to demodulation error

IP_PLL=projection(ampOS_PLL,phOS_PLL,phSF0);
QD_PLL=rejection(ampOS_PLL,phOS_PLL,phSF0);

BI_PLL=vpa(abs(BI0-IP_PLL*3600/ampSF0)); % In-phase BI in dph
BQ_PLL=vpa(abs(BQ0-QD_PLL*3600/ampSF0)); % Quadrature BI in dph

%% ----TIA phase shift----
dphase_TIA=vpa(dphi_TIA); % Total demodulation phase error
% BI_ph_TIA=vpa(dphase_TIA*BQ0) % BI due to demodulation error

IP_TIA=projection(ampOS0,phOS0+dphase_TIA,phSF0);
QD_TIA=rejection(ampOS0,phOS0+dphase_TIA,phSF0);

BI_TIA=vpa(abs(BI0-IP_TIA*3600/ampSF0)); % In-phase BI in dph
BQ_TIA=vpa(abs(BQ0-QD_TIA*3600/ampSF0)); % Quadrature BI in dph

%% ----Total demodulation phase error caused BI-----
BI_ph=sqrt(BI_dw^2+BI_PLL^2+BI_TIA^2)
BQ_ph=sqrt(BQ_dw^2+BQ_PLL^2+BQ_TIA^2)

%% ----Instrument error----
BI_instr=BI0*20*10^-6;
BQ_instr=BQ0*20*10^-6;

%% ----Total BI-----
BI_total=sqrt(BI_dw^2+BI_PLL^2+BI_TIA^2+BI_instr^2)
BQ_total=sqrt(BQ_dw^2+BQ_PLL^2+BQ_TIA^2+BQ_instr^2)
```

References

1. Y. Yang, et al. Measurement of the Nonlinear Elasticity of Doped Bulk-Mode MEMS Resonators, in *Proceedings of Solid-State Sensors, Actuators, Microsystems Workshop*, (2014), pp. 8–12
2. Y. Yang, et al., Nonlinearity of degenerately doped bulk-mode silicon MEMS resonators, J. Microelectromech. Syst. **25**(5), pp. 859–869 (2016)
3. M. Shahmohammadi, H. Fatemi, R. Abdolvand, Nonlinearity reduction in silicon resonators by doping and re-orientation, in *2013 IEEE 26th International Conference on Micro Electro Mechanical Systems (MEMS)*, (2013), pp. 793–796
4. H. Wen, et al. HARPSS-fabricated nano-gap comb-drive for efficient linear actuation of high frequency BAW resonators, in *2016 IEEE 29th International Conference on Micro Electro Mechanical Systems (MEMS)*, (2016), pp. 1014–1017. © 2016 IEEE. Figures and tables reprinted with permission

The manufacturer's authorised representative in the EU is Springer Nature Customer Service Centre GmbH, Europaplatz 3, 69115 Heidelberg, Germany. If you have any concerns regarding our products, please contact ProductSafety@springernature.com

Printed and bound by CPI Group (UK) Ltd, Croydon, CR0 4YY
30/09/2026
02225665-0003